Martin Freligh

The Homoeopathic Pocket Companion

Or, A simplified Abridgment of the Homoeopathic Practice of Medicine

Martin Freligh

The Homoeopathic Pocket Companion
Or, A simplified Abridgment of the Homoeopathic Practice of Medicine

ISBN/EAN: 9783337211493

Printed in Europe, USA, Canada, Australia, Japan

Cover: Foto ©berggeist007 / pixelio.de

More available books at **www.hansebooks.com**

THE
HOMŒOPATHIC
POCKET COMPANION;

OR,

A SIMPLIFIED ABRIDGMENT

OF THE

"Homœopathic Practice of Medicine."

DESIGNED EXPRESSLY FOR THE USE OF FAMILIES AND TRAVELLERS.

BY DR. M. FRELIGH,

AUTHOR OF THE "HOMŒOPATHIC PRACTICE OF MEDICINE," LATE RESIDENT AND VISITING PHYSICIAN TO THE NEW YORK HOMŒOPATHIC DISPENSARY ASSOCIATION, MEMBER OF THE HAHNEMANN ACADEMY, AND OF THE HOMŒOPATHIC MEDICAL SOCIETY OF THE STATE OF NEW YORK.

SEVENTH EDITION

New York:
CHARLES T. HURLBURT.
No. 808 BROADWAY.
1868.

PREFACE.

It is now nearly two years since the publication of the author's "Homœopathic Practice of Medicine." The flattering reception which that work has had throughout the country affords additional evidence of the steady increase of popular confidence in the principles and practice of Homœopathy, and has encouraged the author to contribute still further, if possible, to the universal and permanent adoption of the safest and most successful of the healing arts. The former volume was designed both for professional and non-professional use. Its extensive circulation among the laity, however, has suggested the propriety and utility of issuing a simplified abridgement adapted exclusively to domestic use. Homœopathy being, from the specific, sure, and systematic nature of its treatment, more readily serviceable, in all cases, for non-professional use, than any other method of practice, it is reasonable to believe that a book convenient in size, and describing in brief and familiar terms, the nature, symptoms, and treatment of all ordinary diseases, would be very useful in numerous cases, as, for example, as a pocket companion for private use during travelling, and when the assistance of the physician is scarcely necessary, or when such assistance could not easily be obtained. It is with the hope of furnishing such a complete, simple, and safe manual for family guidance that the present volume is presented to the public. Although an abridgement of the former work, it yet contains all that is essential for this purpose, and only those portions of the other work are omitted which are exclusively historical or theoretical, and therefore of interest solely to the professional reader, as also those more complicated and dangerous forms of diseases which demand the knowledge and skill of the physician and not unfrequently the surgeon. It has been the author's care to avoid everything that would tend to confuse and embarrass the reader, and to bring the work within the comprehension and use of an average degree of intelligence. To prevent confusion and mistake, the administration of the medicines is indicated for every disease, both as to form (whether pillets, powders, or tinctures,) and the dose required.

The alphabetical arrangement is adopted as being the simplest and more convenient, and all technical terms are avoided as far as possible. The author offers his little volume to the public as nothing more than its title declares it to be, and claims for it no more appreciation than its merits deserve.

LIST OF REMEDIES

CONTAINED

IN THIS VOLUME, WITH THEIR ANTIDOTES.

Remedies.	Antidotes.
Acid-muriatic,	Camphor.
Acid-nitric,	Mercurius-sol.
Acid-phos.,	Camphor.
Aconite,	Vinegar.
Agaricus,	Coffea.
Alumina,	Bryonia.
Antimonium-crud.,	Hepar-sulph.
Antimo.-tart.,	Ipecac.
Arnica,	Camphor.
Apocynum-can.,	Bryonia.
Apis-mel,	China.
Arsenicum,	China, Pulsatilla.
Aurum,	China.
Belladonna,	Coffea, Camphor.
Borax,	Cham., Coffea.

1*

LIST OF REMEDIES.

Remedies.	Antidotes.
BRYONIA,	Rhus-tox.
CALCAREA-CARB.,	Acid-nit.
CAMPHOR,	Opium.
CANTHARIDES,	Camphor.
CARB.-VEG.,	Cumphor.
CARBO-ANIMAL,	Arsen., Lachesis, Camph
CAUSTICUM,	Coffea.
CHAMOMILLA,	Coffea.
CICUTA,	Arnica, Camph.
CINA,	Bry., Chin., Ipecac.
CINCHONA or CHINA,	Arsen., Pulsatilla.
COCCULUS,	Camphor.
COFFEA,	Nux-vom.
COLOCYNTH,	Coffea.
CONIUM,	Coffea.
CROCUS,	Acon., Bell., Opii.
CUPRUM,	Ipecac.
DIGITALIS,	Opium.
DROSERA,	Camphor.
DULCAMARA,	Camph., Ipecac., Merc
EUPHRASIA,	Camph., Pulsatilla.
FERRUM,	China.
GRAPHITES,	Arsen., Nux-vom.
HELLEBORE,	Camph.
HEPAR-SULPH.,	Vinegar.
HYOSCIAMUS,	Belladonna.

LIST OF REMEDIES.

Remedies.	Antidotes.
IGNATIA,	*Camphor.*
IPECAC.,	*Arsen.*
JALAPPA,	*Camphor.*
KALI-CARB.,	*Camphor.*
LACHESIS,	*Arsenicum.*
LAUROCERASUS,	*Camphor, Coffea, Ipecac.*
LEDUM,	*Camphor.*
LYCOPODIUM,	*Camph., Puls.*
MERCURIUS-SOL.,	*Sulphur.*
MERC.-CORS.,	*Hepar-sulph., Mezeri.*
MEZERIUM,	*Camph., Mercurius.*
NUX-VOMICA,	*Camphor.*
OPIUM,	*Camphor.*
PETROLEUM,	*Acon., Nux-vom.*
PHOSPHORUS,	*Camphor.*
PLATINA,	*Pulsatilla.*
PULSATILLA,	*Nux-vom.*
RHEUM,	*Camphor.*
RHUS-TOX,	*Bryonia.*
RUTA,	*Camphor.*
SABADILLA,	*Camphor.*
SABINA,	*Camphor.*
SAMBUCUS,	*Camphor.*
SECALE-CORNU,	*Solan.-nigr.*
SEPIA,	*Aconite.*
SILICEA,	*Hepar-sulph.*

Remedies.	Antidotes.
SPIGELIA,	*Camphor.*
SPONGIA,	*Camphor.*
STAPHYSAGRIA,	*Camphor.*
STANNUM,	*Pulsatilla.*
STRAMONIUM,	*Belladonna, Vinegar.*
SULPHUR,	*Merc.-sol., China.*
SYMPHYTUM,	*Merc.-sol., China.*
TEUCRIUM,	*Camphor.*
THUYA,	*Camphor.*
URTICA-URENS,	—
VERATRUM,	—
ZINCUM.	*Camph., Ignatia.*

Tinctures used for External Application.

Tincture of *Arnica.*   Tincture of *Ruta.*
" *Symphytum.*   " *Urtica-urens.*
" *Cantharides.*   " *Camphor.*
Aqua Ammonia.

The Tinctures of *Arnica*, *Symphytum*, and *Ruta*, are used for Strains, Sprains, Bruises, Contusions, and Wounds.

The Tincture of *Urtica-urens* for Burns and Scalds.

The Tincture of *Cantharides* for Chilblains and some forms of Erysipelas.

The Tincture of *Camphor* and *Aqua Ammonia* for the bites and stings of insects.

DIETETICS.

The success of Homœopathic treatment, in many instances, depends as much upon the proper observance of dietetic rules as upon the selection of the drug. For if the diet is not in accordance with the ends to be accomplished by the medicines, the latter must fall far short of accomplishing the object for which they were administered, and this depends as much (and frequently more) upon the influence of improper articles of food and condiments in interfering with the action of the medicines, by partially or fully antidoting their effects, than upon their pernicious effects upon the system singly considered.

Under the several diseases, I have generally indicated the appropriate diet; but it may not be amiss to note under a general head the articles considered objectionable or positively injurious during the use of homœopathic medicines, viz.:

beef, pork, fish, and other meats, salted, or salted and smoked, or pickled in vinegar. Fish not having scales, as catfish, eels, lobsters, crabs, and clams. Highly seasoned dishes, with pepper, spices, or aromatics. Cakes or other pastry prepared with soda, cream of tartar or saleratus; spiced, aromatic, or flavored confectionery.

Vegetables and fruits, celery and spinage, water cresses, lettuce, peppergrass, parsley, radishes and parsnips, pineapples, sour oranges, lemons, bananas, and the ordinary sour fruits and nuts of an oily and aromatic nature.

* * * Distilled and malt liquors; mead, soda, and mineral waters; lemon, root, soda, and spruce beers. Avoid also Allopathic drugs of every name and form; and the use of mustard paste, the fumes of vinegars, colognes, camphor, hartshorn, aromatic vinegar, smelling salts and the ordinary pungents; also carefully guard against the inhaling of carbonic acid or sulphurous gas which frequently escapes from grates, furnaces and stoves, in which anthracite coal or the ordinary charcoal is burned.

THE DOSE

AND FORM OF MEDICINE.

It will be observed that I have in most instances named, or prescribed the proper dose of medicine to be given, but where I have not, the following rules must be observed, viz: When pillets are used, one or two may be placed on the tongue and allowed to dissolve, followed by a draught of water; or five or six may be dissolved in a tumbler half full of pure cold water, well stirred, and from a tea-spoonful, to a table-spoonful, given at a dose, as the case may seem to require, and whether the patient be a child or an adult. When tinctures are used, mix from one to three drops in a similar quantity of water, stir it well, and give it in spoonful doses similarly. When triturations (powders) are used, give as much as will lay on the point of a penknife-blade, either dry on the tongue, or dissolved in water similar to the pillets.

THE HOMŒOPATHIC POCKET COMPANION.

ABSCESS.

An abscess is "a collection of pus in the cellular membrane, or in the viscera, or in the bones, preceded by inflammation."

TREATMENT.—The remedies usually employed in the treatment of abscesses are *Arsenicum, Assafœt., Belladonna, Bryonia, Chamomilla, Hepar-sulph., Ledum, Mezerium, Phosphorus, Pulsatilla,* and *Sulphur,* viz:

When the abscess is attended with severe burning pains, and threatens to become gangrenous, or when there is great debility, *Arsenicum* every four or six hours.

When the abscess discharges a colorless, watery

pus; pains severe on contact; and sensitiveness of the adjoining parts, *Assafœt.*, as above.

But when attended with a pressing, burning and stinging in the abscess; and a floculent cheesy pus is discharged, *Belladonna* three times a day.

When the tumor is either red or pale, and attended with a tensive pain, *Bryonia*, every three or four hours, until relief is obtained.

When the abscess is of the *fibrous parts*, and of *tendons*, and *ligaments;* or when it arises from the abuse of mercury, *Mezerium*, morning and evening.

When the abscess is of a *Lymphatic* character, with fistulous openings, surrounded with callous edges and discharging a fœtid unhealthy pus, and all the symptoms are worse at night, *Phosphorus*, morning and evening.

When there is a tendency to bleed, with stinging and cutting pains, or an itching and burning in the surrounding parts, *Pulsatilla*, every four or six hours.

When the abscess is located in the *axillæ* (arm pit) or the parotid gland (of the cheek and angle of the jaw); the swelling painful to the touch, and if the discharge is of a bloody serous character, *Rhus-tox* two or three times a day.

When the abscess originates from a diseased bone, *Arsen.*, *Calc-carb.*, *Assafœt.* and *Aurum*, to be given twice a day. Give the remedy first named for one week; should the patient improve-continue it, but if there is not a material improvement give the next similarly, and so continue them until a cure is effected.

For LUMBAR ABSCESS, an abscess in the lumbar region, (small of the back), or, for PSOAS ABSCESS (located between the psoas muscles, which lie posterior to the kidneys), *Calcarea-carb*, *Kali-carb*, *Ferrum*, and *Silicea*, given three times a day, each remedy to be continued from three days to a week.

For Abscess of the Liver, (*Hepatic Abscess*,) *Belladonna* and *Silicea*.

When an abscess appears of a decidedly scrofulous character; or for an abscess in a scrofulous subject, *Arsen.*, *Calc-phos*, *Ferrum*. The remedies should be given (as they are here arranged) morning, noon, and night, and continued until a manifest improvement or cure is effected.

ADMINISTRATION.—Give five or six pillets at a dose; but if the medicine is in the form of a tincture, mix two or three drops in a tumbler about one third filled with pure cold water; stir well, and give from a tea-spoonful to a table-spoonful at a dose.

EXTERNAL APPLICATIONS.—There is no objection to the use of poultices, such as ground flax seed, slippery elm, bread and milk, and such like, during the forming stage of an abscess, but they should never be used in case of a fully matured and discharging abscess. Washing with tepid water and castile soap, and the use of simple dressings, are best.

DIET AND REGIMEN.—In many instances a nourishing diet is strictly demanded, together with such other means as will tend to invigorate the system. But when an abscess occurs in a robust subject, with a predisposition to inflammatory diseases, the diet should be restricted to the simplest kind.

ABDOMEN.

ABDOMEN, (from *abdo*, to hide, because it hides the viscera,) the belly, "the largest cavity in the body," separated from the chest (*Thorax*) by the midriff (*diaphragm*) and bounded inferiorly by the bones of the pelvis.

ABDOMINAL DROPSY.—(*Ascites.*) Vide Dropsy of the Abdomen.

ABDOMINAL PAINS.—Vide Pains in the Abdomen.

AFTER-PAINS.

After-pains appear necessary in the contraction of the womb, and are sometimes quite severe and distressing, when coagulums are formed. They are not so severe during the first confinement, but seem to increase at every subsequent accouchment.

TREATMENT.—Give *Arnica* immediately after delivery, to soothe and remove the soreness.

If the pains continue severe, of a drawing or tensive character, with a pale face and a degree of coldness, use *Pulsatilla* every two hours until relieved.

If the pains are of a pinching, spasmodic character, in the abdomen, and attended with discharges of coagulated blood, *Chamomilla* as above.

When the pains are of a contractive character and mostly in the back and hips, *Nux-vomica* every two hours until easier.

For after-pains in those who have had several children; when the patient is rather weak, with a disposition to flooding, *Secale*, as in the case of the last remedy.

When the pain is attended with great sensitiveness, weariness of the limbs, and extreme nervousness, *Coffea* every hour until the patient is better.

ADMINISTRATION.—Mix one or two drops, or dissolve six or eight pillets in six tea-spoonfuls of water, and give a tea-spoonful at a dose. Keep the patient perfectly quiet and well bandaged.

AGUE IN THE BREAST.

This painful condition of the breasts is generally caused by cold or some obstruction, and requires *Aconite, Bryonia, Pulsatilla, Chamomilla, Calcarea-carb.,* and *Phosphorus*.

When head-ache and fever are present, give *Aconite* every hour or two until the fever abates.

When the breast is swollen and inflamed, and there is a suppression of milk, use *Bryonia* every three or four hours until relieved.

When the breasts are swollen, and attended with

a painful pressure, or a sticking sensation, and the discharge of milk is thin and acrid, *Pulsatilla* every three or four hours.

When there is a hard, indurated condition of the breast; soreness of the nipples, which are painful to the touch; give *Chamomilla*, every four hours.

For a bruised, sore pain in the breast, and soreness of the nipples, *Calcarea-carb.* twice a day.

When there is hardness, erysepelatous inflammation, swelling and stitches, accompanied with burning and stinging; *Phosphorus*, morning and evening.

When the breasts are swollen and painful and threaten suppuration, give *Hepar-sulph* every two or three hours.

ADMINISTRATION.—If the medicine is in the form of pillets, take five or six at a dose. But if it is a tincture, mix two or three drops in a tumbler half full of water, stir well and take a dessert-spoonful at a dose.

APOPLEXY.

Apoplexy is defined as a sudden loss or suspension of the sensorial functions and voluntary motion. It frequently comes on suddenly without any warning symptom or assignable cause. But it is generally preceded by dizziness, a dull pain and degree of weight in the head, throbbing of the temporal arteries, ringing in the ears, sparkling before the eyes, confused ideas, drowsiness and heavy sleep.

TREATMENT.—When the face is flushed; the eyes injected, and there is a violent beating and throbbing of the temporal arteries, give *Aconite* and *Belladonna* in alternation every four hours.

When the attack is incomplete and the patient is conscious of suffering, and evinces signs of sore pain in the head; and if attended with numbness of the limbs, full pulse, and an occasional twitching of the extremities, *Aconite* and *Nux-vomica*, as above.

When the patient appears as if in a profound sleep, face of a natural appearance, pulse rather slow and not increased in volume, *Opium* every two hours until the patient rouses from the stupor.

When caused by a blow or fall, *Arnica*.

When caused by suppressed menstruation, from cold, *Pulsatilla* and *Sulphur;* then, *Opium* or *Belladonna*.

When caused by repelled eruptions, *Sulph. Ipecac.*

When caused by fright, *Ignatia*, *Pulsatilla*, *Artemisia*.

ADMINISTRATION.—Apply three or four pillets to the tongue of the patient every four hours, as there is generally a difficulty of swallowing.

EXTERNAL APPLICATIONS.—Cold water to the head and warm or even stimulating applications to the feet and legs.

APPETITE.

LOSS OF APPETITE, (ANOREXIA.)—This condition generally arises from a derangement of the stomach and digestive organs, but sometimes appears as symptomatic to other affections.

TREATMENT.—For loss of appetite, indifference to food; flat, watery, or bitter taste; use *Chamomilla* an hour before eating.

If the loss of appetite is attended with sickness of the stomach, use *Antim-crud.* a few minutes before eating.

If it is attended with a sweet, insipid taste, and sickness of the stomach, *Ipecac.*, as above.

If it is attended with a putrid taste, and a desire for acid drinks, use *Bryonia* in the same manner.

If attended with sickness of the stomach, with great aversion to milk and warm food, *Ignatia*, two or three times a day.

When there is a salt, or metallic taste; pressure in the pit of the stomach, with nausea, and gulping up of sour water after eating, *Lachesis*, three times a day, from half an hour to an hour before eating.

When loss of appetite is associated with a putrid, bitter, or bad taste, sour raisings and frequent hiccough; pressure, tension, and cramp-like pains in the pit of the stomach; and the bowels are constipated, use *Nux-vomica*, three times a day, until the symptoms are removed.

For loss of appetite, with a bad taste in the mouth; when the throat appears narrowed; pressure and tension when swallowing, and pressure in the pit of the stomach, *Pulsatilla* three times a day, an hour before eating.

For complete loss of appetite, with bitter taste, and dryness of the throat, *Rhus-tox*, as above.

For a loss of appetite, attended with a putrid or sour taste, and constant thirst, *Sulphur* three times a day before eating.

ADMINISTRATION.—Dissolve ten or twelve pillets, or if the medicines are tinctures, mix two or three drops in a tumbler from half to two thirds full of water, stir well, and give from a tea-spoonful to a table-spoonful at a dose.

DIET AND REGIMEN.—The diet should be unirritating, easily digestible, and of the most agreeable kind to the patient, but in accordance with homœopathic rules.

APPETITE VORACIOUS.

A voracious appetite or insatiable hunger is frequently produced by worms, particularly the Tape Worm, and it not unfrequently occurs during convalescence from severe diseases, such as Typhoid and other fevers, and sometimes during pregnancy.

TREATMENT.—For excessive hunger, particularly at night, a "yearning for dainties;" great thirst, but drinking little at a time, with a desire for wine and sour drinks, *China* three times a day, an hour before eating.

For voracious appetite, attended with nausea

after eating, and colic or colic and diarrhœa, give *Colocynthis* similarly.

For a voracious hunger, from a sickish feeling of emptiness, weakness of digestion, and frequent eructations, *Natrum-mur.* and *Nux-vom.*, the former in the morning and the latter at night.

For great appetite, with increased thirst, *Stramonium* or *Sulphur*, an hour before meals.

For an insatiable hunger when the stomach is full, *Staphysagria* as above.

When there is a great desire for sweet things, *Lycopodium* every morning.

When there is a desire for earthy substances, such as chalk, lime, &c., *Nit-acid.* once a day.

If the hunger is attended with unquenchable thirst for cold drinks; the tongue dry, and dark or yellow-coated, *Veratrum* before breakfast and dinner.

When it is caused by the common intestinal worm, *Cina* three times a day.

When caused by the Tape Worm, *Mercur.*, *Graph.*, *Stramonium*.

ADMINISTRATION.—Give five or six pillets at a

dose, or if the medicines are in the form of powders, give as much as can be held on the point of a penknife blade at a dose; if they are tinctures, mix two or three drops in a gill of water, stir well, and give from a tea-spoonful to a table-spoonful at a dose.

DIET AND REGIMEN.—The diet should be nutritious and of easy digestion, and restricted to a moderate quantity.

APTHAE, (*Thrush or Sprue.*)

This disease is almost exclusively confined to children. It consists in white flakes, of the appearance of magnesia, chalk, or curdled milk, which appear on the tongue, inside of the mouth, gums, fauces, and frequently extend to the stomach and bowels.

TREATMENT.—*Sulphur* is the best remedy in this affection while the thrush appears white.

Mercury.—When it changes to an ash, bluish, or grayish color.

Ferrum.—When it changes to a yellowish brown, which is frequently the case when the attack is severe.

ADMINISTRATION.—Dissolve ten or twelve pillets in about a wine-glassful of water, and give a small sized tea-spoonful every four or six hours, or if powders are used, place a small quantity on the tongue once or twice a day.

ASTHMA (*Phthisic.*)

Asthma is a paroxysmal affection of the respiratory organs, characterized by great difficulty of breathing, tightness and oppression of the chest, and a sense of impending suffocation.

TREATMENT.—For oppressed, laborious breathing, gasping for breath, with the mouth open and face flushed, give *Aconite* every fifteen or twenty minutes, until the patient is relieved.

When the breathing is laborious and irregular; short and hurried and then slow; face changing from pale to redness, and bloated, *Belladonna* every half hour, or hour, until relief is obtained.

When the breathing is spasmodic, chest oppressed, and palpitation of the heart, *Cocculus* every half hour, until relieved; then extend the time to two hours.

When the breathing is spasmodic, attended with

spasmodic cough, or the breathing is short, quick, and attended with wheezing, and the patient is almost suffocated, use *Cuprum* every half hour until relieved.

When the breathing is short and panting, and the chest feels as if contracted, *Ipecac.* every half hour, or hour, until the patient is better.

When the tightness and difficult breathing come on in paroxysms; face hot, with dizziness of the head; use *Lobelia* every hour until the paroxysms cease.

When there is loud breathing, and panting; or with a feeling of suffocation, *Phosphorus* every half hour.

When the difficulty appears more in the throat than the chest, and is attended with a hoarse cough; use *Petroleum* every hour or two, until it is relieved.

ADMINISTRATION.—Dissolve ten or twelve pillets in a tumbler half full of cold water; or if the medicine is a tincture, mix two or three drops in a like amount of water, stir well, and give to an adult a table-spoonful at a dose, and to a child a tea-spoonful.

It will be perceived that the remedies are recommended at short intervals, because the attack is generally very severe and exceedingly threatening.

But in chronic Asthma, once a day, or every second day, is sufficiently often to take the remedies.

DIET AND REGIMEN.—The diet should be light, unirritating, easy to digest, and such as will not interfere with the medicines employed. Avoid night air and a humid atmosphere.

BILIOUSNESS.

The term bilious is generally used to express diseases and conditions arising from a derangement of the liver, and is applied to colic, diarrhœas, fevers, etc. But we are frequently called upon to prescribe for a biliary derangement, producing languor, heaviness, headache, loss of appetite, a disinclination to physical and mental exertion, and other associate symptoms, independent of the more formidable diseases above mentioned.

TREATMENT.—When there is headache, with a yellow appearance around the nose and mouth, and the bowels are constipated, give *Nux-vomica* and *Pulsatilla;* the latter morning and noon, and the former at night.

But when there is a sluggishness and a disposition to sleep during the day, face yellow and scurvy, loss of appetite, and the bowels are irregular, *Mercurius* night and morning.

When there is headache, a confused state of the mind, flat and bitter taste, thirst and pain in the region of the liver, *China* three times a day.

When the skin appears yellow in spots, with loss of appetite and restlessness at night, *Colchicum* as above.

But when it produces a general yellowness of the skin; loss of appetite, with nausea; pain in the bowels, and a frequent desire to urinate, give *Arsenicum* two or three times a day.

Chamomilla is a good remedy when there is a dull, oppressive headache, bitter taste, nausea, and uneasiness in the pit of the stomach.

For a mere bilious condition, attended with languor, heaviness of the head, yellow skin, variable appetite, etc., *Pulsatilla* and *Mercurius* in alternation, one in the morning and the other in the evening, will generally prove sufficient; but if the bowels are constipated, use *Nux-vomica* at night.

ADMINISTRATION.—Give to an adult five or six

pillets at a dose, to a child two or three. If the medicines are in liquid form, mix two or three drops in about a gill of water, stir it well, and give from a tea-spoonful to a table-spoonful at a dose.

DIET.—The diet ought to be light, so as not to tax the digestive organs, for dyspepsia is prone to set in upon biliary derangement.

BILIOUS COLIC.

This is a severe and dangerous form of colic, characterized by severe pains in the stomach and bowels, attended with bilious vomiting and obstinate constipation, and is generally caused by an excess of acrid bile.

TREATMENT.—In case of severe bilious colic, attended with pain of a twisting or cutting character; sickness of the stomach, and obstinate constipation, give *Aconite* and *Nux-vomica* in alternation, every fifteen or twenty minutes.

When the pain is writhing and twisting about the navel, give *Bryonia* every half hour.

When the pain is above the navel, and so severe as almost to arrest breathing, and attended with constant sickness of the stomach, give *Cocculus*, as above.

When the pains are severe, with a violent bearing down and straining, with nausea, or vomiting of bile and mucus; give *Belladonna* every fifteen or twenty minutes until relief is obtained.

External applications, such as hot fomentations—by applying cloths wet in hot water, and wrung sufficiently dry to prevent soiling the clothes, will generally afford great relief.

Another important feature in the treatment is, to move the bowels as soon as possible, by means of injections.

Allopathy invariably prescribes cathartics; never use them, as the stomach is too irritable to retain them; and they always aggravate the difficulty by their irritative effects.

ADMINISTRATION.—Give five or six pillets at a dose, or if the medicine is in liquid form, mix two or three drops in a gill of water, stir it well, and give a large-sized tea-spoonful at a dose.

BITES AND STINGS OF INSECTS AND SNAKES.

For the pain and inflammation arising from the sting of a bee, apply to the part the ordinary *Spirits of Hartshorn.*

For the bite of spiders and ants, wash the part with *Spirits of Camphor*, or diluted *Tincture of Arnica;* one part of *Arnica* to three of water.

For the bite or sting of the small fly, called the gnat, or for the bite of mosquitoes, *Camphor* or *Arnica*, as above.

For the bite of that most poisonous serpent, the Copper-head, the internal administration and the external application to the part bitten, of a decoction of the *Broad-leafed Plantain* is advised, and I have known it used successfully in two or three instances.

BLEEDING FROM THE NOSE.
(*Epistaxis.*)

Bleeding from the nose generally occurs among the young, those of full habit and not yet arrived at maturity; and more frequently in males than females, particularly after menstruation becomes established.

TREATMENT.—When it occurs in persons of full habit, who are subject to head-ache, dizziness, flushed face, and heating and throbbing of the temporal arteries, give *Aconite* and *Belladonna*, in alternation, every fifteen or twenty minutes, until it is arrested.

But when it attacks the weak and delicate, with a pale face, and general weakness, *China* and *Ferrum* every half hour.

When it occurs at night, give *Rhus-tox*, every fifteen or twenty minutes, until it arrests it.

When in the morning, *Bryonia* in the same manner.

When caused by a blow or an injury, *Arnica* similarly.

When caused by the free use of intoxicating drinks, *Nux-vom.* and *Lachesis*, separately or in alternation, every half hour until it ceases.

When caused by a violent exertion in lifting or straining, *Rhus-tox*, or *Carbo-veg.*

To correct a disposition to bleeding from the nose, give *Aconite, Sulphur, Sepia,* and *Lycopodium*, a dose of each every second day.

DIET.—The diet should be in accordance with homœopathic rules during the administration of medicines.

BLEEDING FROM THE LUNGS.

(*Hæmoptysis.*)

Bleeding from the lungs is generally preceded by pain, or a degree of fullness in the chest, some heat in the throat, and a sweet, insipid, saltish taste in the mouth; and is characterized by coughing up a florid, frothy blood, which last distinguishes it from hæmorrhage from the stomach, which is dark in appearance.

TREATMENT.—When bleeding of the lungs occurs in a full, plethoric habit, or in a young, vigorous person, give *Aconite* and *Belladonna*, in alternation, every fifteen or twenty minutes, until it ceases.

When the bleeding is preceded by oppression in the chest; some difficulty in breathing; fetid breath, and bad taste in the mouth; a tickling sensation in the throat, exciting a dry, hacking cough; and the blood raised by coughing is bright red and frothy, intermixed with small clots, or coagulums, give *Arnica* every half hour.

When the breathing is laborious, with hurried expirations; or oppressed breathing, in consequence of stitches in the chest; dry cough, producing retching, and an expectoration of pure blood, or bloody mucus, give *Bryonia*, every fifteen or twenty minutes.

When there is an asthmatic shortness of breath, and a continuous, uninterrupted cough which occasions spitting of blood, give *Ipecac.* every half hour.

When there is a loss of appetite; bitter or sweetish taste; nausea; the heart beats quick and laborious; the pulse small and slow; face pale; lips livid; sore feeling in the chest; cough, with a bloody expectoration, or florid blood, use *Digitalis* as above.

When the bleeding occurs in a person of mild temperament, or in one disposed to melancholy, *Pulsatilla* every half hour or hour.

And *Pulsatilla* more especially if there are asthmatic symptoms, with cough, relieved by sitting up; occasional stitches in the side; alternate chills and flashes of heat; and the attacks occur towards evening, or during the night.

When the bleeding from the lungs is caused by the drying up of an old ulcer: the patient is weak and somewhat emaciated; some cough, with constant hawking up of pure blood, use *Rhus-tox** every fifteen or twenty minutes.

---

* I have successfully treated several attacks of this dis-

When the bleeding occurs in a person addicted to intemperance, *Nux-vomica* or *Lachesis*, every half hour or hour.

When caused by a blow, fall, or the inhalation of irritating particles of dust or other substances, *Arnica*, as above.

When caused by metastasis from Gout or Rheumatism; or by the drying up of an old ulcer, *Bryonia*, or *Rhus-tox.*

When caused by suppressed menstruation, treat for the primary difficulty, establish normal action there, and the bleeding from the lungs will cease.

When caused by a violent exertion, use *Rhus-tox.*

ADMINISTRATION.—Dissolve ten or twelve pillets of the medicines in a tumbler half full of iced water, or if Tinctures are used, mix three or four drops in a like quantity of water, stir it well, and give a dessert-spoonful as directed above, until a favorable impression is made; then extend the

---

ease in a person whose condition was similar to that expressed in the above paragraph, with *Rhus-tox;* it was employed after the other drugs which were more strictly indicated had failed, and afforded almost immediate relief; subsequent attacks in the same patient were also almost immediately arrested by it.

time to every hour, or to two, three or four as may be necessary.

DIET AND REGIMEN.—If food be necessary before the bleeding is arrested, or during the intervals, it must be of the simplest character, and eaten cold; such as cold, soft boiled rice, and cold gruel or ponade; the drinks must consist of iced water. Keep the patient perfectly quiet and preserve a tolerably low temperature in the apartment.

BLEEDING FROM THE STOMACH.

(*Hæmatemesis.*)

Blood raised from the stomach can readily be distinguished from that of the lungs by the appearance of the blood, which is dark and grumous, and frequently mixed with the ordinary contents of the stomach; it is also generally preceded by a sense of weight, by pain, or a degree of uneasiness in the pit of the stomach, and is unattended with cough.

TREATMENT.—*Aconite* and *Belladonna* are in this, as well as in the other forms of hæmorrhage, the best remedies we possess for cases occurring in those of full habit. Give them in alternation every hour until the bleeding ceases.

Hyosciamus and *Crocus* are also good remedies, the former when there is that peculiar congested state, with a red, bloated or bluish face; livid lips and sickness of the stomach. The latter when there is a degree of heaviness in the head and vertigo, and when the blood raised is dark and viscid.

When it occurs during low fevers, *Arsenicum* every two or three hours.

When it occurs during pregnancy, *Nux-vomica* every hour; and if this does not arrest it, give *Kreosote* every hour.

If it is caused by a blow or mechanical injury, *Arnica*.

If caused by a suppression of the menses, *Pulsatilla* and *Sulphur*, in alternation, every four hours.

ADMINISTRATION.—Give the remedies as in the other forms of hæmorrhage.

DIET AND REGIMEN.—The most simple, bland, mucilaginous and unirritating articles of food, and these should be cold. Iced water as a drink, and the occasional swallowing of small pieces of ice is beneficial, and the apartments should be of a low temperature.

BLEEDING FROM THE URETHRA.

(*Hæmaturia.*)

Bloody urine is generally the result of an injury, either from a fall or calculous concretions in the kidneys and bladder, or from their lodgement in the urethra. It is sometimes occasioned by the injudicious use of acrid and stimulating diuretics, such as *Spanish fly*, *Turpentine*, etc.

TREATMENT.—When it is caused by a fall, bruise, or any kind of mechanical injury, *Arnica;* mix five or ten drops of the tincture in a tumbler half full of cold water, and give a table-spoonful every two or three hours.

When it is attended with much pain and burning, and a constant desire to urinate, *Arsenicum* as above. If that does not arrest it, give *Cantharides* similarly.

When it is attended with stitches in the parts, and the urine deposits a bloody sediment, or is mixed with blood, *Calc.-carb.* every two or three hours until relief is obtained.

When it is attended with a cutting pain in the bladder, or a bleeding from the urethra, without pain, *Lycopodium*, as above.

When there is a cutting, painful urinating, attended with bloody discharges, *Phosphorus* every hour until relief is obtained.

ADMINISTRATION.—Give the remedies as above directed, and extend the time according to the improvement.

DIET AND REGIMEN as in the other forms of hæmorrhages.

BLEEDING FROM THE WOMB.
(*Menorrhagia.*)

This consists in an immoderate flow, or hæmorrhage from the womb, and is generally attended with pain in the back, loins, hips, and abdomen.

TREATMENT.—If an immediate flow of the menses occurs in a person of full plethoric habit, give *Aconite* every two or three hours.

When it is attended with much bearing down pain, the blood discharged appears bright red, and contains clots, *Belladonna* every hour or two until it is arrested.

But when it is attended with cutting pain in the abdomen, pain in the small of the back, and the blood discharged is mostly in clots, use *Nux-vomica*, as above.

When the discharge is thick and offensive, or corrosive, and attended with a burning and soreness, use *Phosphorus.*

When the pain is more of a griping, colicky character, and the discharge is in clots, *Chamomilla* every hour until it brings relief.

If hæmorrhage takes place during pregnancy, and threatens miscarriage, use *Secale-cornu.*

If the discharge is dark, and appears like partly decomposed blood; with pale face and stupefaction of the senses, *Stramonium* and *Cicuta,* in alternation every hour.

When it is caused by a violent strain or exertion, *Sulph.-acid, Ruta,* or *Rhus-tox,* as above.

ADMINISTRATION.—Dissolve ten or twelve pillets, or two or three drops of the medicine, if in tincture, in about a gill of water, and give a table-spoonful at a dose. Keep the patient in a recumbent posture, allow no warm drinks, and keep the room rather cold than otherwise.

BLINDNESS. (*Amaurosis.*)

A partial or complete loss of sight is generally the result of a paralysis of the optic nerves, particularly when there is no deformity of the eye from mechanical or acute diseases. It is frequently caused by violent exertions, but generally arises from pressure on some part of the eye from turgescence, or from debilitating losses, and sometimes from uterine irritation. Lawrence, in his work on the eye, cites the case of a Jewess who became completely blind during three successive pregnancies; a very similar case occurred in my own practice.

TREATMENT.—For a degree of indistinctness of sight, with a sensation of pressure and weakness in the eyes, *Aurum*.

When the blindness appears as though a gauze were drawn before the eyes, *Ruta*.

When there is dimness of sight, with inflammation, redness and swelling; the dimness increased by warmth or exercise, and the patient rather of a phlegmatic temperament, *Pulsatilla*.

When it is attended with some swelling of the

lower eyelids; the eyes appear watery; and objects appear green or watery, *Digitalis.*

When the attack comes on suddenly, without any assignable cause, *Nux-vom., Rhus-tox,* and *Veratum.*

When it depends upon a turgid state of the blood vessels from plethora or convulsions, *Belladonna, Hyosciamus.*

When caused by violent exertion, lifting or straining, *Rhus-tox.*

For mere weakness of sight, caused by excessive reading, fine sewing, or embroidery, *Ruta.*

When caused by an injury, *Arnica.*

When it arises from habitual intemperance, *Nux-vom., Ignatia, Lachesis.*

When it occurs during pregnancy, *Bell., Nux-vom., Pulsatilla.*

When it is caused by debility, or is the result of debilitating losses, *China, Ferrum.*

When caused by sexual abuses, *Phos., Staphysagria, Sepia.*

For night blindness, *Belladonna.*

For blindness during the day, *Sulp.* and *Silicea.*

ADMINISTRATION.—If the medicine be in the form of tincture, mix one or two drops, or if in the form of powder, as much as will lie on the point of a penknife blade, in a tumbler half-full of pure cold water; stir it well, and to an adult patient give a dessert-spoonful every six or eight hours; to a child give a tea-spoonful at a dose. When pillets are used place two or three on the tongue; allow them to dissolve, and swallow them with a draught of water. If the case is chronic, and has been of long standing, one dose of the medicine daily is sufficient.

DIET AND REGIMEN according to Homœopathic restriction.

BOILS OR BILES.

A boil is a circumscribed, prominent, deep red, inflammatory swelling, exceedingly painful, which terminates slowly by suppuration, and frequently produces considerable constitutional disturbance, particularly in irritable persons, and, in children, sometimes causes spasms.

TREATMENT.—When it is attended with fever and headache, *Aconite*.

BOILS OR BILES.

If the boil is much inflamed, painful, and slow to mature, *Belladonna*.

When the boil is large and threatens to become carbunculous, *Arsenicum* and *Belladonna*.

When it is small, and there is a rough unhealthy skin generally, give *Sulphur* and *Rhus-tox* in alternation.

When there is a disposition to boils, give *Lycopod.*, *Phos.*, and *Sulphur:* each twice a week for three or four weeks.

ADMINISTRATION.—When the boil is large and produces considerable suffering, give the remedy selected every three or four hours, until relief is obtained; then extend the time to every six or eight hours.

EXTERNAL APPLICATIONS.—Simple poultices of bread and milk, or slippery elm, will tend very materially to mitigate the suffering, and favor suppuration.

I have seen the greatest advantage result from the external use of the *Tinct. of Arnica*, by mixing a few drops, say ten or twenty, in a gill of water, and applying it to the part, by means of old fine linen. It is particularly useful when applied during the forming stage.

BRONCHITIS.

By this term we understand an inflammation of the air passages of the lungs, which is characterized by a slight cough, hoarseness, tightness and oppression of the chest, and sometimes by a rattling wheezing, as if the air were passing through a small aperture, or were obstructed by mucus.

TREATMENT.—For an attack of acute Bronchitis give *Aconite* every two or three hours, until the symptoms are somewhat subdued, then *Bryonia* similarly, for at least three or four administrations.

If the lungs continue congested, and the face is flushed, give *Belladonna* similarly until relief ensue.

BRONCHITIS CHRONIC.

This form is far more frequent than the acute, and is considered by some rather fashionable than otherwise. There are those, however, who are predisposed to it by physical conformation; particularly asthmatic persons.

TREATMENT.—If there are any feverish symptoms, *Aconite* and *Bryonia* should be given in alternation every four hours, the latter particularly if there is soreness of the chest and a mucous expectoration.

When it is attended with hoarseness in the morning, and an expectoration of greenish, saltish, or white catarrhal mucus, *Phosphorus*, morning and evening.

DIET AND REGIMEN.—The diet should be light and unirritating; take moderate exercise and avoid night air, and the impure air of crowded apartments.

BRUISES AND CONTUSIONS.

For bruises or contusions produced by a fall, blow, or otherwise, there is nothing better to prevent soreness, swelling, and discoloration of the part, than the external application of diluted *Tinct. of Arnica*, say one part of *Arnica* to five or six parts of water. Bathe the part frequently with it or keep it applied by means of old fine linen wet in the liquid.

Should the bruise or contusion be of the side, or

of the chest, or affect the head, give five or six pillets of *Arnica* every two or three hours.

But if fever or inflammation should set in after the use of the above remedy, give *Aconite* every three or four hours until the fever abates.

BUBO.

Bubo is an inflammation and enlargement of one or mere of the inguinal, and sometimes, of the external iliac, lymphatic glands. When it arises from mere local irritation or from cold, it is termed *sympathetic bubo;* when from the absorption of syphilitic virus, *syphilitic* or *pestilential bubo.*

TREATMENT.—For syphilitic bubo, *Merc.-cors., Iodine, Mezeri.;* and emollient applications, such as *Slippery-Elm* and *Flaxseed.*

ADMINISTRATION.—Give *Merc.-cors.* morning and evening, in from three to five grain doses of the first trituration for one week. The *Iodine* and *Mezeri.* may be taken in alternation every six or eight hours.

Poultice the part with *Slippery-Elm* or *Flaxseed.*

DIET.—The diet must be very simple, and principally vegetable.

BURNS AND SCALDS.

The danger to be apprehended from burns and scalds depends of course altogether upon the depth or extent of the injury. In numerous cases the injury has been most dreadfully aggravated by the use of one or more of the popular remedies recommended by those who ought to know better.

The only treatment necessary, and the most reliable under all circumstances, is the application of *Urtica-urens*, by mixing forty or fifty drops of the tincture in a half-pint of water of an ordinary temperature, and keeping it constantly applied to the part by means of a piece of old fine linen immersed therein. This must be continued until the inflammation is removed, which will be indicated by the departure of pain, and the perfectly white appearance of the injured part. Then keep the part covered with old, fine linen, lightly spread with clean, fresh mutton-tallow, or simple cerate, in order to lubricate it and protect it from the air.

If the burn is an internal one, caused by the inhalation of steam, or scalding liquid, give a teaspoonful of the mixture every two or three hours, until the suffering is mitigated.

CARBUNCLE. (*Anthrax.*)

A carbuncle is a hard circumscribed inflammatory swelling, like a boil, which sometimes forms on the cheek, neck, or back, and in many instances becomes gangrenous or mortifies. The discharge is generally extremely offensive, of a thin, bloody looking matter, which exudes from beneath and around a dark core; in some cases there are two, three, or more openings.

TREATMENT.—When there is pain of a beating and throbbing character, fever, pain in the head, and tenderness of the part, *Aconite* and *Belladonna* in alternation every three hours.

When the carbuncle discharges a thin, bloody, offensive matter, and the patient complains of a burning pain, *Arsenicum* every three or four hours until the symptoms subside.

When there is an inability to sleep at night; a degree of pressure and tension about the ulcer, with a peculiar burning and smarting, and the discharge is a thin, bloody pus, *Carbo-veg.* every four hours.

When the ulcer appears dark, and as if it were mortifying; the discharge exceedingly offensive;

with a disposition to sloughing, and a want of sensibility in the part, *Kreosote* three times a day; and apply to the ulcer a wash composed of ten or twelve drops of *Kreosote* to a gill of water, in order to cleanse and arouse some action in the part.

Diet.—The diet should be rather liberal, but compatible with homœopathic treatment.

CATARACT

Is a species of blindness arising from an opacity of the crystalline lens, or its capsule, which prevents the rays of light passing to the retina.

It generally commences with a disturbed sight, as if motes or particles of dust were in the eye, or floating before it, which are termed *muscœ volitantes;* and, as the opacity increases, the sight becomes less perfect, until it is entirely lost.

Treatment.—It may be well to observe that cataracts are not always successfully reached by medicines, and in case of failure, can only be remedied by an operation; but I have cured cases by homœopathic remedies, and have two cases now improving finely under similar treatment.

When the sight is obscured as if by a gauze web, or mist before the eye; some itching of the nose, with dry coryza, *Causticum, Silex, Baryta-carb.*

If the sight is obscured when in the open air, or dark clouds, motes or specks appear before the eyes, *Pulsatilla, Conium-mac., Lachesis.*

When the eyes are prone to be sore, especially in scrofulous persons; the sight weakened and imperfect; black spots appear passing before the eyes, or luminous vibrations, *Phosphorus.*

When there is a weakness of sight; inability to recognise anything with distinctness; black motes appear before the eyes; itching of the lids, or dryness of the eye, with a burning sensation, *Hepar-sulph., Sulphur, Calc.-carb.*

When the cataract is caused by an injury to the head or eye, 1st. *Arnica;* 2d. *Conium.*

When it appears secondary to *Syphilis*, or in a person of syphilitic taint, *Mercurius, Nit.-acid.*

Administration.—The disease under consideration is not one demanding profuse medication, nor the rapid administration of remedies; their use once a day, or every second day, will prove quite sufficient. When two or more remedies are named to the same condition, administer them in order as they are placed, giving the one named first for a reasonable length of time, and continuing it if an improvement is manifest; but should there be no material improvement, or the improvement from the first not increase, then give the next in order similarly, and so on.

Diet and Regimen.—The diet must be compatible with the remedy employed, and every thing tending to strain or weaken the eye, strictly avoided.

CATARRH.

This is an affection of the mucous membranes of the nostrils and frontal sinuses. The nose appears stopped in many instances, with loss of smell, or a constant disagreeable odor and taste; pain at the root of the nose, or a rattling and cracking sensation, and dull, heavy headache. The secretion sometimes passes down the posterior nostrils to the throat, causing a constant scraping or hawking up of a white, greenish, or yellow and bloody expectoration.

Treatment.—When the discharge is thin and acrid, and attended with burning in the nose, and frequent sneezing, or a pain at the root of the nose, *Arsenicum* once a day.

When there is a stoppage of the nose, and the nose is painful and sore to the touch, use *Aurum* every morning.

When it is attended with scaly formations in the nostrils, and loss of smell, *Hepar-sulph.* once a day.

When there is swelling of the entire nose, and irritation of the nostrils, attended with a copious discharge of a thin, watery character, use *Mercurius* morning and evening.

But if it is attended with a fetid smell, and a discharge of matter, bloody or greenish in appearance, or the nose is swollen, with a loss of smell, *Phosphorus* every morning or evening until it is better.

If the bones of the nose are affected, use *Aurum* in alternation with the *Phosphorus*.

ADMINISTRATION.—Give from three to five pillets at a dose, or if the medicine is in powder, give as much as will lay on the point of a pen-knife blade.

DIET AND REGIMEN.—In accordance with homœopathic restrictions during the administration of redies.

CHANCRES.

Chancres are small, ash-colored ulcerations on the genitals. Sometimes they are no larger than a pin's head, and sometimes they appear as mere abrasions of the skin. They are caused by syphilitic virus.

TREATMENT.—Give the first trituration of *Mercurius-cors.* in from five to six grain doses, morning and evening, and touch the part with *Caustic* (*Argentum-Nit.*), which, with proper perseverance and cleanliness, will soon effect a cure.

The remedy should be continued as above for at least two or three weeks, so that we may be assured that the disease is entirely eradicated from the system.

DIET AND REGIMEN.—The diet must be restricted to a plain, simple, unirritating course, and all kinds of stimulating drinks strictly avoided.

CHICKEN-POX. (*Varicella.*)

This disease was supposed by many to be a species of Small-Pox, and described as such, and continues to go under various cognomens. It is undoubtedly one of the same family, but of an exceedingly mild character, and seldom attended with much fever. The eruption appears very irregular, pustules appearing, forming and scabbing at the same time.

TREATMENT.—If there is any fever, give *Aconite* every three or four hours until it subsides.

When it is attended with pain in the back, and drawing pains in the limbs, *Bryonia* every four hours until these symptoms pass off.

When there is pain in the head : oppression of the chest; and a creeping chilly feeling, one or two doses of *Pulsatilla.*

If the patient is a child, and symptoms of difficulty of the head appear, give *Belladonna* and *Pulsatilla* every two hours, until relief is obtained. The latter drug will generally mitigate the symptoms very materially if given in the commencement of the attack; and will almost act as a complete specific to the disease.

CHILBLAINS.

Chilblains are the effect of an inflammation arising from cold. In their mildest form, they are attended with redness of the skin, some swelling.

burning and itching. When more violent, the swelling, burning and itching become much worse; the color of the part is a dark red or even blue; sometimes blisters arise upon the tumor, which burst and leave excoriations, which may run into an indolent ulcer.

TREATMENT.—When there is burning, biting, stinging and itching, and not much swelling or discoloration, *Agaricus*.

When there is inflammation, with burning and itching, *Arsenicum* first, every morning and evening; should that not have the desired effect in a reasonable time, give *China* or *Nit.-acid* similarly.

If there is much swelling of a dark red or blueish color, give *Arnica* and *Belladonna* in alternation, every four or six hours.

When they are large and exceedingly painful, give *Hepar-sulphur* or *Phosphoric acid*, or *Sepia;* a dose two or three times a day.

EXTERNAL APPLICATIONS.—Much relief will be found from the application of diluted *Tinct. Cantharides*, if there is much burning and stinging, and a disposition to form blisters; and diluted *Tinct.* of *Arnica* when there is a peculiar burning and stinging, with swelling.

CHOLERA. (*Cholera Asiatic.*)

This is one of the most fearful diseases with which we are called upon to contend. Not because it is beyond the reach of Homœopathic treatment (for it is the only treatment that can be relied on;) but because there are so many popular remedies generally resorted to, such as cholera mixtures, etc., before the physician is called, that it is a hard matter to obtain the specific effect of remedies on a dying person nearly drugged to death with opium, camphor, hot drop and calomel. Such drugs can never be employed in a single case, with a rational hope of surviving the attack.

An attack of Cholera is sometimes very sudden and extremely prostrating, but it is generally preceded by diarrhœa, sickness of the stomach and spasms of the muscles of the chest, which are soon followed by distressing vomiting and purging of a substance much resembling rice water, with cramps of the muscles of the chest, abdomen and extremities; rapid prostration and sinking; an anxious expression; livid lips with a blueish tinge of the entire surface, pulse nearly or quite imperceptible; and an icy coldness of the extremities.

CHOLERA.

PREVENTION.—Acting upon the principle that "an ounce of prevention is better than a pound of cure," I suggest the necessity of a strict adherence to the following: during the prevalence of Cholera avoid every thing that may tend to act as a predisposing cause; such as fear; irregularities and excesses of any kind; avoid all food, fruit and drinks that tend to relax or irritate the stomach and bowels; wear flannel under-clothing, and sleep between flannel sheets; preserve a cheerfulness of mind, and attend to business as usual; make no great or sudden change from the usual mode of living, as the change itself will act prejudicially, and observe temperance and moderation in all things.

TREATMENT.—Immediately at the commencement of the disease, place the patient in bed in a recumbent position, and cover him with warm flannel; pay particular attention to the warmth of his extremities, and give drop doses of the ordinary *Tinct.* of *Camphor* in a tea-spoonful of iced water every five or ten minutes until a warm glow is excited. Then extend the time to every fifteen or twenty minutes, until a moderate perspiration is produced and the breathing becomes freer, and keep the patient perfectly quiet for six or eight hours, or until his recovery is beyond all doubt.

But if the disease proceeds to the second stage, give *Veratrum* every ten or fifteen minutes so long as there are cramps and violent spasmodic action of the stomach and bowels. But if the cramps are more of the chest, with great difficulty of breathing and the lips and face are livid (blueish), give *Cuprum* as above. But if the cramps are more of the calves of the legs, or the patient is threatened with stupor; and particularly if he is an elderly person, *Secale Cornutum* is advised; keep the extremities constantly warm by all possible means.

If the disease passes to the third or collapse stage, give *Arsenicum* and *Veratrum* in alternation, every ten or fifteen minutes. *Arsenicum* is indicated in almost every stage in consequence of the burning thirst, vomiting and marked prostration. With *Camphor* alone, I have been signally successful.

ADMINISTRATION.—If the pillets or powders of *Arsenicum*, *Veratrum* and *Cuprum* are used, administer them as above, by placing three or four pillets or a small powder on the tongue. The use of *Camphor* I have sufficiently directed.

DIET AND REGIMEN.—During convalescence great care is requisite. The diet should be very mild, consisting of simple mucilaginous articles, at first, then of mild broths and simple toast. Avoid every kind of exercise until returning strength warrants an effort.

CHOLERA INFANTUM.

This is a disease which, as the term implies, is "peculiar to children." It generally commences with sickness of the stomach and vomiting of an acid smelling fluid; diarrhœa with thin watery discharges, resembling in many instances soap suds and exceedingly offensive; an anxious expression of countenance; eyes sunken, with a dark streak beneath them, and the skin pale and rather cold. The little sufferer soon becomes exceedingly restless, throwing itself from one side of the couch to the other, and there is intense thirst for cold drinks which are rejected as soon as swallowed.

TREATMENT.—At the commencement, when the vomiting and purging is of a bilious character, give two or three pillets of *Ipecac.* every half hour or hour.

But when the discharges are thin; and the thirst is very great; pale skin and prostration, give *Arsenicum* as above.

CHOLERA INFANTUM.

When in addition to the last condition, the mouth is parched, the tongue dark, the lips black or cracked, *Veratrum* in alternation with *Arsenicum*.

When the disease has affected the brain, and the child rolls its head on the pillow; and the face is flushed, give *Belladonna* every hour or two. But if the face is pale and the pupils of the eyes are dilated, give *Digitalis*.

Should symptoms of fever set in upon reaction, moderate them by an occasional pillet or two of *Aconite*.

DIET AND REGIMEN.—During the attack, give nothing but cold water in very small quantities; anything else will excite vomiting; and when the stomach will retain anything more do not feed the little creature with brandy and milk punch, as is frequently done by nurses and others who ought to know better; and thereby cause inflammation of the brain to terminate in effusion. Properly prepared arrow-root, corn starch, and farina is all sufficient.

CHOLERA MORBUS.

Cholera Morbus occurs generally during summer. It is a severe vomiting and purging, attended with griping pains in the stomach and bowels.

TREATMENT.—At the commencement of the disease there is generally some feverish excitement which *Aconite* will frequently arrest at once. But if the disease continues, with a vomiting and purging of a bilious character, give *Ipecac.* every fifteen or twenty minutes.

But if the discharge from the stomach and bowels is thin and watery, and the patient is covered with perspiration, give *Antim.-tart.*

When the vomiting is very severe; and the discharges from the bowels are watery, bilious, or dark appearing; the extremities cold and the patient's strength prostrated, give *Arsenicum.*

If the pains are very severe, sharp, or cutting, and particularly if they are more in the vicinity of the navel; and for cramps of the muscles of the legs, *Verat.*

When the discharges are watery and almost colorless, *Colocynth.*

When the disease is caused by cold drinks, or sudden changes of temperature; and attended with bilious stools and vomiting, *Dulcamara*.

Administer the remedies selected every fifteen or twenty minutes until they produce a decided change for the better, then extend the time as the patient improves.

DIET AND REGIMEN.—Plain unirritating articles of food, at first such as thin gruel of oat-meal, toast and barley water, arrow-root, rice, etc. When all the symptoms are subdued, a more liberal diet may be indulged in, such as beef tea, mutton broth and toast.

CHRONIC RHEUMATISM.

This form of Rheumatism is generally characterized by pain and lameness of a muscle or joint, and although exceedingly annoying at times, is never associated with the general symptoms of excited pulse, fever, etc., as in the acute form.

TREATMENT.—When it affects the joints generally, *Colchicum*.

When mostly located in the shoulder and elbow, *Pulsatilla* and *Bryonia* in alternation.

CHRONIC RHEUMATISM.

When the back is mostly affected, *Nux-vomica* morning and evening, and if that does not cure, give *Lycopodium* in the same manner.

When it affects the knee joints; with stitches, particularly when moving them, *Bryonia* every four hours.

When it is in the hip joint; particularly the left hip; or if there is a degree of uneasiness or heaviness in the lower limbs, *Sulphur*.

When the pains are principally in the knees, and are of a lancinating or jerking character, or for inflammatory swelling of the knee, *Pulsatilla*.

When the pains are of an aching bruised feeling, and especially if they are worse at night, *Rhus-tox*, every evening.

But when the pains are mostly of the muscles, attended with numbness, *Plumbum* twice a day.

And in addition to the above remedies, the application of dry hot cotton batting to the affected part, will be found in many instances very serviceable.

By the term Clap is understood a purulent discharge from the Urethra, of a greenish yellow color, attended with heat, some swelling and inflammation of the part, burning and scalding when urinating, and painful erections.

The cause of this disease is too well known to require any explanation.

TREATMENT.—Subdue the general inflammation with a few doses of *Aconite;* then the local with *Mercurius-cors.* and *Sabina* or *Balsam-copaiva.* Should the discharge continue after most of the local inflammation has subsided, make use of a solution of *Nitrate* of *Silver*, 10 grs. to an ounce of water, as an injection, and use it every evening until the discharge ceases.

ADMINISTRATION.—Use from three to five grains of the first trituration of *Mercurius-cors.* at a dose every three or four hours, and from two to three drops of the *Tinct.* of *Sabina* in the same way, or they may be used in alternation, until the heat and inflammation pass off.

DIET AND REGIMEN.—The diet must be light

and unirritating. Avoid all kinds of pepper and spicy seasoning, and also the use of stimulating drinks. Mucilaginous and gummy drinks are best. Take as little exercise as possible.

COLDS.

Cold is unquestionably the cause of most of our acute inflammatory diseases. But there are so many minor symptoms produced by exposure to cold, and embraced within the common saying of " taking cold," that reference to such symptoms appears very necessary in a popular work like this.

TREATMENT.—For a Cold affecting the head, if attended with fever, give *Aconite* every three or four hours until the feverish symptoms subside.

If it produces a stoppage of the nose, with irritation, give *Nux-vomica* and *Dulcamara*. If it produces a running at the nose, *Sulphur*.

If it produces severe headache give *Aconite;* if that does not relieve, give *Belladonna* and *Bryonia*, in alternation every two or three hours.

If it produces a general chilliness or shivering, with alternate flashes of heat, *Pulsatilla* every two hours, until these symptoms are removed.

When it produces severe pain in the back, pains of a drawing character, and extending up to between the shoulders, *Dulcamara* and *Bryonia* in alternation, every three or four hours.

For severe pains in the back caused by cold; the pains of a lame, aching character, and preventing stooping, *Nux-vom.*, *Lycopodium* ; the latter morning and noon, the former at night.

For severe aching and drawing pains in the limbs, caused by cold, *Bryonia* and *Rhus-tox* in alternation, every three or four hours. When the pains are worse at night, give *Aconite* or *Rhus-tox ;* should those remedies not relieve, give *Lachesis* and *Carbo-veg*.

For sore throat, from cold, *Aconite* and *Belladonna* in alternation every three or four hours.

For a suppression of the menses from cold, give *Pulsatilla* in the morning and *Sulphur* at night, until they have the desired effect.

For the effects of cold from getting wet, at first give *Dulcamara*, and if that does not entirely relieve, give *Belladonna* and *Pulsatilla* in alternation.

For great sensitiveness to changes of weather, *Carbo-veg.*, *Lachesis* and *Sulphur*. Give the remedy first-named once a day for two or three days, then the others in rotation similarly.

ADMINISTRATION.—When pillets are used, simply lay one or two of them on the tongue and allow them to dissolve. The powder may be used as previously directed, and the Tincture, by mixing two or three drops of the medicine selected in a tumbler half full of water, and a dessert-spoonful taken at a dose, if the patient is an adult; but if a child, a tea-spoonful is sufficient.

DIET AND REGIMEN.—Do not be led astray by the old absurd idea of "feeding a cold and starving a fever," but let the diet be light and in strict accordance with homœopathic treatment.

COLIC.

The term colic is generally applied to all pains in the abdomen, and particularly when the bowels are constipated. Dr. Cullen defines it as "pain of the abdomen, particularly around the umbilicus, (*navel*,) attended with vomiting and costiveness."

TREATMENT.—When the attack of colic is produced by a congested state of the bowels, which is characterized by a weak pulse, and the extremities inclined to be cold, and also a degree of numbness, give *Aconite* and *Belladonna*, in alternation, every twenty minutes or half hour, until relief is obtained.

But when the pain is below the navel, and is of a grasping, twisting character, give *Belladonna* every fifteen or twenty minutes until it is relieved.

When it is more of a bilious colic, attended with vomiting of bile, the tongue is yellow and coated, *Nux-vomica* every half hour.

When it is more of a wind colic, and the pains extend more from the left to the right side, *Carbo-veg.* as above.

For wind colic, with most of the suffering above the navel, and inclining to the right side, *Cocculus* every half hour.

For flatulent or wind colic, producing pains in different parts of the abdomen, *Chamomilla*.

Chamomilla is also the best remedy for the ordinary wind colic of infants.

Colocynth is also an admirable remedy, when there is an occasional emission of wind, which is followed by violent cutting pains in the bowels.

Secale-cornu, for the colic attending menstruation.

ADMINISTRATION.—Dissolve eight or ten pillets, or a small powder of the medicine selected, in ten or twelve tea-spoonfuls of cold water; stir it well, and administer as above. If Tinctures are used, mix from one to three drops in a tumbler one-third or half full of water, and give tea-spoonful doses until the patient is relieved.

CONGESTION.

The term Congestion is usually used when the vessels are over distended, and the motion of the circulating fluid is slow.

CONGESTION OF THE HEAD.

Congestion of the Head is generally characterized by beating and throbbing of the temporal arteries; giddiness; a heavy, dull feeling; partial blindness when stooping or suddenly turning around, or a degree of confusion and indistinctness.

TREATMENT.—When this condition occurs in a full, plethoric habit, give *Aconite* and *Belladonna* in alternation, every three hours until it is relieved.

When it is attended with pain of one side of the head and the bowels are constipated, *Nux-vomica* three times a day.

When attended with stupefaction, and partial loss of consciousness, *Opium* every hour until the stupefaction passes off.

When it is caused by a fall or blow, *Arnica* every hour or two. If by fright or fear, *Opium* in the same manner.

When it is the result of debility, *China* and *Ferrum*, one in the morning and the other at night.

ADMINISTRATION.—Mix two or three drops of the medicine indicated in a tumbler half full of water, and give from a tea-spoonful to a table-spoonful at a time. If pillets or powders are used, dissolve six or eight pillets, or a small powder, in a similar amount of water and give similarly.

CONGESTION OF THE LUNGS.

This condition of the lungs is characterized by pain, a degree of fullness, tightness, and oppression of the chest, with difficult breathing, palpitation of the heart, faintness, disturbed sleep, and sometimes cough.

TREATMENT.—*Aconite* and *Belladonna*, in alternation every hour until the patient is relieved, particularly if he is a robust person.

When it is attended with much difficulty in breathing, and palpitation of the heart, *Digitalis* and *Pulsatilla* in alternation.

When a congested state of the chest is caused by a mere trifling exertion, and is attended with faintness and difficult breathing, *Spongia*.

ADMINISTRATION.—Mix two or three drops of the Tincture, or ten or twelve pillets, or a small powder, in a tumbler partially full of water and give a dessert-spoonful, if the attack is severe, every half hour until it is relieved. But for chronic congestion, or a mere disposition to a congested state, once a day is sufficient.

CORNS. (*Clavus.*)

Corns are remarkably unpleasant, painful companions, and are the result of pride in most instances, from wearing boots and shoes far smaller than the dimensions of the feet honestly demand. They are simply hardened portions of cuticle, (scarf skin,) produced by pressure, and have a sort of core or thorn, that can be worked out, something in appearance to a barley córn, whence the name. (Hooper.) But I must confess that I have never yet seen anything about them much resembling a barley corn, but have supposed the name was derived from the hard, dense, semi-transparency, like horn.

TREATMENT.—When they are inflamed and painful, wash them frequently with diluted *Arnica Tincture*, and apply some lint, or wrap up the part in a piece of old soft linen, wet in the above tincture; or carefully shave them down until the skin appears of a natural thickness, and apply *Arnica plaster*, which has certainly proved very efficacious in a great many instances.

CONSTIPATION OR COSTIVENESS.

Costiveness is generally a condition which depends upon an inactivity of the liver; but with many it is constitutional. It should, however, be attended to, and the habit corrected if possible.

TREATMENT.—*Mercurius* and *Nux-vomica* are very good remedies, particularly when there are any prominent symptoms of biliary derangement.

Give the former in the morning and the latter at night.

When it occurs in the summer, and there is a feeling of chilliness, some headache, and particularly if the person is subject to Rheumatism, give three or four pillets of *Bryonia* at a dose, once or twice a day.

When it is attended with a sensation of heaviness in the abdomen, dry mouth, loss of appetite, headache, and occasional drowsiness, give *Opium* morning and evening.

For obstinate constipation, attended with a pressing sensation in the stomach, *Lachesis* once or twice a day.

For constipation of infants at the breast, *Bryonia*, *Nux-vom.* or *Allum.* Dissolve five or six pillets of either in a wine-glass full of water, and give a small sized tea-spoonful three times a day, until it is remedied.

For constipation of old persons, *Opium* and *Phosphorus;* give the former first, a dose once a day for three or four days, and if it is ineffectual, give the latter similarly.

For constipation of drunkards, give *Lachesis* in the morning, and *Nux-vom.* at night.

DIET.—The diet should be of a relaxing character, and calculated to assist the action of the remedies.

CONSUMPTION. (*Phthisis Pulmonalis.*)

The term consumption is generally applied to a tuberculated disease of the lungs, which has proved more destructive of human life than any in the long catalogue of ills which frail humanity is heir to. In its treatment too much attention has been paid to the condition of the lungs at an expense of a proper regard to the peculiar condition of the system, upon which these symptoms depend; for most truly is it, (the condition of the lungs), in the language of Williams, in his "Principles of Pathology," "a mere fractional part of a great constitutional derangement."

For a full and complete history of the disease, its prevention, pathology, and cure, vide Freligh's "Homœopathic Practice," page 207.

PREVENTION.—The best way to prevent a development of consumption where a predisposition exists, is to invigorate the system by good nourishing diet, regular exercise in fresh air and equal temperatures, regular rest, cheerful company, and warm clothing, frequent ablutions with cold water sufficient to invigorate the skin and preserve perfect cleanliness, and avoid all kinds of excesses.

Give *Ferrum, Calcarea-carb., Calcarea-phos., Kali-carb., Silicea* and *Sulphur*. These remedies must be used of the first trituration, and in doses sufficient (say from five to ten grains) to invigorate the system by increasing the solid constituents, extractive matters and salts. *Ferrum* stands most prominent and should be given first. I would advise one dose a day for a week, then the other remedies in succession in the same way. Should the patient be inclined to lose flesh give *Arsenicum* first, in similar doses every morning, and *Ferrum* at night; continue thus until there is an evident increase of flesh and strength.

CONVULSIONS, OR FITS.

The remote causes of Fits in children are sometimes veiled in obscurity, but they are very frequently caused by the irritation from teething, or worms.

TREATMENT.—The best course to pursue is to place the child immediately in a warm bath, or its feet and legs in a warm bath with a little mustard in it; after using the bath a sufficient length of time (say five minutes), wrap the child in warm flannel, and administer *Belladonna*, by placing two or three pillets, or a small quantity of the powder on the tongue; place the child's head and shoulders in an elevated position, and use gentle friction of the extremities.

But should the convulsion continue, with violent motions of the limbs, twitching of the muscles of the face, and foaming at the mouth, give *Hyosciamus* as above, every fifteen or twenty minutes.

Should convulsive jerkings continue, the eyes partly open, with redness of one cheek, give *Chamomilla* in the same manner.

If the convulsions are caused by worms, give two or three pillets of *Cina*, and should they continue, give either *Belladonna* or *Hyosciamus*, according to their indications above.

When caused by fright, give two or three pillets of *Opium*; should that not arrest them, give *Stramonium* or *Hyosciamus* similarly.

LOCAL REMEDIES.—Should the above remedies fail, after using them a reasonable length of time, move the bowels by means of a tolerably stimulating injection.

COUGH.

Cough is most generally a symptom, particularly in diseases of the throat and lungs, but it very frequently appears independent of any marked disease, and simply as the result of cold or some other irritating cause.

TREATMENT.—For a dry, hacking cough, give *Aconite*, *Nux-vomica* or *Bryonia*.

For a dry cough at night, disappearing when sitting up in bed, *Pulsatilla*.

For a severe cough, setting in immediately after going to bed at night, *Aconite*.

For a hollow cough at night and early in the morning, *Causticum*.

For a loose, rattling cough, *Tart.-emetic*.

For cough, with expectoration of tough mucus, *Dulcamara*.

For cough, with expectoration of pus, *Calc.-carb.* If the pus is streaked with blood, give *Pulsatilla* in alternation with the *Calcarea*.

When the cough is attended with quite a bloody expectoration, *Rhus-tox*. If it is attended with a yellow expectoration, give *Ignatia*. When the expectoration is a whitish mucus, *Cina*. When the expectoration is muddy, give *Phosphorus*.

When the cough is worse in the morning give *Bryonia*, but when it is worse at night give *Belladonna*. Should three or. four administrations fail to effect a cure, give *Hyosciamus*.

For coughs in children, caused by teething, give *Chamomilla* every three or four hours.

. For coughs following measles, *Ignatia* three times a day.

ADMINISTRATION.—Give two or three pillets of the remedy indicated every two, three, or four hours, according to the urgency of the cough; but if the cough is very severe and almost constant, the remedy may be given every fifteen or twenty minutes, until relief is obtained.

DIET AND REGIMEN.—Avoid all irritating and indigestible articles of food, and stimulating drinks. Take moderate exercise in pure wholesome air, and do not go out at night, while troubled with a cough.

CRAMPS IN THE LIMBS.

The principal remedies for cramps in the limbs are *Colocynth, Rhus-tox, Veratrum, Secale-cornu, Hyosciamus, Calcarea-carb., Bryonia* and *Lycopodium.*

When the cramps are in the hips, give *Colocynth* or *Rhus-tox* three times a day.

For cramps in the calves of the legs, use *Veratrum* morning and evening. Should that not have the desired effect, use *Secale-cornu,* or *Hyosciamus,* similarly.

For cramps in the feet and toes, use *Calcarea-carb., Bryonia,* or *Lycopodium,* once or twice a day.

For cramps during pregnancy, vide *Pregnancy.*

ADMINISTRATION.— Mix two or three drops of the medicine in a tumbler one-third full of pure cold water, or if pillets are used, dissolve ten or twelve of them in a like quantity of water and give a dessert-spoonful at a dose, as directed above, unless the attack is severe and continuous, when the remedy should be given every half hour or hour until relief is obtained.

Diet.—Such as will not interfere with the action of the medicines.

CRAMPS IN THE STOMACH.

Cramp, or spasmodic pain in the stomach, requires *Aconite*, *Belladonna*, *Calcarea-carb.*, *Carbo-veg.*, *China*, *Lycopodium*, *Rhus-tox* and *Veratrum*.

When it is severe and attended with feverish excitement, give *Aconite* and *Belladonna* in alternation every five or ten minutes until it is relieved.

When it is more of a pressure or contracting pain, *Bryonia* three times a day.

When it is very violent, attended with gulping up wind and tasting the food, give *Calcarea-carb.* every hour until it is relieved.

For cramp of the stomach, with a sense of pressure, *Carbo-veg.* every half hour or hour.

When it occurs particularly after a meal, *Nux vomica* every half hour until it is removed.

When it is more of an intense pain, *Lycopodium* in the same manner.

For an uneasy pressure in the pit of the stomach,

as if it were swollen, *Rhus-tox.* every hour or two until it has the desired effect.

When the cramp is attended with violent pressing pain in the pit of the stomach; or a sharp, cutting, piercing pain, *Veratrum* as above.

DIET.—Persons who are subject to cramps or pains in the stomach should carefully avoid all kinds of acid and indigestible fruits, pastry, fresh bread and salt meats.

CROUP.

Croup is an inflammation of the mucous membrane of the trachea or windpipe, and sometimes ascends quite high up in the throat, (*larynx.*)

It is characterized by hoarseness when crying; cough of a peculiar hoarse, hollow sound; drawing a long breath, with a stridulous or crowing noise. The cough becomes more troublesome and shrill; the breathing more and more difficult; the face flushed and swollen, and each breath adds to the agitated and anxious expression of the little sufferer. The severe symptoms generally come in paroxysms which threaten suffocation, when the little patient throws its head back to straighten the air passages, in order to relieve the suffocation. It is one of those insidious and flattering diseases, that require the closest watching, and the most energetic treatment.

TREATMENT.—During the inflammatory stage, give *Aconite* every half hour; in most instances this will arrest it.

But should it continue and pass into the second stage, which is characterized by the peculiar hoarse, hollow, suffocative cough and great difficulty in breathing, give *Spongia* every fifteen or twenty minutes; or it may be given every fifteen minutes in alternation with *Hepar-Sulphur.*

When the breathing appears obstructed by an accumulation of mucus, *Causticum* every half hour.

When it appears to be of a spasmodic character, *Hyosciam.* as above.

Kali-bichrom., and *Bromine*, are very popular remedies with some of our most successful practitioners; the former, when the cough is worse in the morning, or immediately on waking; and for violent wheezing and panting; or wheezing and rattling in the chest during sleep. The latter remedy is considered by some a specific and admissible in every stage of croup. I have thus used it upon the suggestion of Dr. Curtis, of this city, with the most decided success. I used it in from two to five drop doses of the tincture, of strength sufficient to retain considerable of its color, (say a yellowish color,) every fifteen minutes at first, then every half hour, and as my little patient improved, every two or three hours. In several instances, two or three administrations removed every symptom.

The other remedies should be given in pillets, two or three placed on the tongue; or in powders similarly, as very frequently a great difficulty in swallowing exists.

DIET AND REGIMEN.—When the child can swallow and evince a desire for nourishment, there is nothing better than that which it receives from its mother; in the absence of which, thin toast or cracker-water may be used to allay the thirst. I am decidedly opposed to arrow root, gum arabic, and such mucilaginous articles as are generally used; for they certainly tend to mix to some extent with the secretion of the throat, adhere, and so increase the difficulty both in breathing and swallowing.

The room should be kept of a moderate warmth; the child by all means secured from all exposure to varying temperatures.

CRYING OF INFANTS.

Mothers and nurses seem to have such a decided preference for pins instead of tapes, for securing the clothing of infants, that it is always necessary to be sure that the crying is not caused by a stray pin.

If not, then resort may be had to the following: *Coffea*, if there is feverish heat, restlessness, and continual crying. *Chamomilla*, if the infant draws and bends its body and limbs, as if in pain; and particularly if the stools are greenish.

Belladonna, when the abdomen is distended. *Rheum*, if it is attended with looseness of the bowels, stools rather yellow and of a sour smell.

Ipecac.—When there is a colicky uneasiness, diarrhœic stools of a fermented appearance, and especially if attended with sickness of the stomach.

Jalap.—When attended with colicky pains and large watery evacuations from the bowels.

ADMINISTRATION.—Dissolve five or six pillets of the medicine indicated by the symptoms, in teaspoonfuls of cold water, stir it well, and give a tea-spoonful every half hour or hour until the child is relieved; or give one or two pillets at a dose.

DIET.—Care of course should be taken that nothing be given in the form of nourishment which is of a flatulent or griping character.

Apply warmth to the bowels, and keep the child quiet and in rather a flexed position, or "*curled-up,*" as the old ladies say.

DELIRIUM TREMENS. (*Mania-a-Potu.*)

This is a variety of mental derangement which is caused by the excessive use of intoxicating drinks, and is characterized by an uneasy, unquiet state, continued watchfulness, cool skin, loquacious delirium, illusions, and almost a constant tremor. The patient is sometimes very merry, at others feels sad, and sees and hears strange things; or fancies that the devil is lurking about the room, watching an opportunity to seize him; or he is busily engaged in killing rats, mice and snakes; and not unfrequently imagines himself transformed into an inferior animal or some article of furniture.

TREATMENT.—The principal indication in the treatment of this disease is to produce sleep and quietude, for which *Aconite* should be given at short intervals, if the attack occurs in a person of full habit; *Belladonna*, when the face is flushed, the patient is wild and furious, and attempts acts of self-violence.

Nux-vomica, when he appears in a vexed mood, and is disposed to quarrel.

Opium.—For a wild and staring expression, the patient grasps at imaginary things, or has visions; the face pale, and covered with perspiration; the pulse rather slow; inability to sleep and a disposition to commit suicide.

Hyosciamus.—When the patient is sleepless and stares suddenly, as if frightened; complains of head-ache; his eyes are red, and there is thirst, with difficulty in swallowing; wishes to fight, kicks, stamps, and strikes violently; trembling of the hands and arms, and coldness of the feet.

When there is a constant change of feeling from a vexed mood to laughing, singing, dancing, and praying, *Stramonium*.

ADMINISTRATION.—Give the remedies indicated every half hour or hour until they have the desired effect. I have generally used the Tinctures in this disease, by mixing from three to five drops in a tumbler half full of water and giving a table-spoonful at a dose.

DIARRHŒA.

Diarrhœa is a purging or mere looseness of the bowels. The discharge varies in character, and may be bilious, mucous, watery, or bloody, depending to a great extent upon the cause.

TREATMENT.—For mere looseness of the bowels, painless diarrhœa, give *Ferrum* or *China*, three times a day; the latter more particularly if the food passes unchanged.

When it is attended with pain, *Nux-vom.* and *Colocynth;* the latter in the morning, and the former at night.

When the discharges are decidedly bilious, give *Mercurius* in the morning, and *Nux-vom.* at night.

If the diarrhœa is attended with a bearing-down pain and straining, *Belladonna* every four hours.

For mucous diarrhœa, use *Pulsatilla* morning and noon, and *Nux-vom.* at night.

If the discharges are thin and watery, and attended with sickness of the stomach, *Antimony* every three or four hours.

When caused by cold, *Dulcamara* in the morning, and *Nux-vom.* at night.

When caused by sour fruit or drinks, *Arsenicum* or *Lachesis*, every three or four hours.

When diarrhœa follows Measles, *Pulsatilla* as above.

When it is secondary to Scarlet Fever, *Belladonna*, similarly as above.

When caused by teething, *Chamomilla* two or three times a day.

When it occurs during pregnancy, *Nux-vom.*, *Phos.*, *Sepia*, or *Sulph.*

For nightly diarrhœa, *Phosphorus*, morning and evening.

ADMINISTRATION.—When more than one remedy is named and not specially directed in alternation with another, give the first one; if that does not cure, give the next, and so on until a cure is effected.

DIET.—Avoid everything of a relaxing or loosening nature.

DIZZINESS. (*Vertigo.*)

This unpleasant symptom generally depends upon a deranged state of the stomach, and congestion of the brain; and is almost a constant companion of plethora.

TREATMENT.—When it attends plethora, (fullness of habit,) give *Aconite* and *Belladonna*, in alternation, every six or eight hours, until it is fully relieved.

When it depends upon a deranged state of the stomach, *Nux-vom.* three times a day until it passes off.

When it comes on upon lying down, *Rhus-tox* every evening, until it is remedied.

When it is caused by suppressed eruptions or the drying up of old ulcers, *Sulphur* and *Lachesis* in alternation, every three or four hours, until it is relieved.

When caused by riding in a carriage, *Cocculus* or *Petroleum*, once a day.

When it is caused by a night's debauch, *Nux-vomica* every hour or two until it is relieved.

ADMINISTRATION.—Give two or three pillets at a time, followed by a draught of water; or dissolve ten or twelve in as many tea-spoonfuls of water, and give a tea-spoonful at a dose.

DROPSY. (*Hydrops.*)

By the term dropsy, we understand a preternatural collection of serous or watery fluid in the cellular substance, or in different cavities of the body.

CAUSES.—The causes of dropsy are various, such as cold, mechanical obstructions, inflammation of particular organs, suppressed perspiration and habitual discharges, repelled eruptions, debility, scirrhus of the liver, the continued use of arsenic in intermittent fevers, the intemperate use of intoxicating liquors; it is also frequently secondary to Measles and Scarlet Fever.

TREATMENT.—During the first appearance of dropsical swellings, if there are febrile or inflammatory symptoms, give *Aconite* every four or six hours until the fever abates.

If the swelling is principally confined to the feet, and occurs during pregnancy, *Lachesis* once a day.

When the legs and feet are swollen, the bowels constipated, the urine scanty and high-colored, loss of appetite, metallic taste and thirst, *Mercurius* three times a day.

When there is a very general swelling, the skin sallow, dry cough, difficult breathing, the extremities feel cold, thirst, urine scanty and high-colored, *Arsenicum* every four or six hours.

When the feet swell during the day, and the swelling diminishes at night, *Bryonia* in the morning and *Sulphur* at night.

When the dropsical swelling succeeds Scarlet Fever, *Apocynum-can.* Drop from three to five drops of the Tincture in a tumbler two-thirds full of cold water, stir it well, and give a large sized tea-spoonful every two or three hours until it produces a manifest improvement; then extend the time to two or three hours. It scarcely ever fails.

When the dropsical effusion depends upon disease of the heart, use *Digitalis* three times a day.

When there is a general swelling and the abdomen is much distended, the urine is passed in diminished quantities, and the other remedies have failed, give one drop of *Apis-mel*, two or three times a day; or, if the medicine is in the form of powders, give of the third trituration about as much as will lie on the point of a penknife, as above, and continue it until there is a marked improvement.

For dropsy arising from debility, give *China*.

When it is caused by mercury, *Hepar-sulph.*

When caused by sulphur, *Pulsatilla.*

When caused by quinine, *Pulsatilla, Arsenicum, Canabis.*

ADMINISTRATION.—Give the drug selected, every four, six, or eight hours, according to the severity of the symptoms. If Tinctures are used, mix from one to three drops in a tumbler half or two thirds full of water, stir it well, and give from a tea-spoonful to a table-spoonful at a time. If pillets are used, dissolve from ten to twelve in a similar amount of water, and administer similarly.

DIET.—The diet ought to be light, unirritating, and of easy digestion, and the amount of drink restricted in quantity.

DROPSY OF THE ABDOMEN OR BELLY.
(*Ascites.*)

This is a swelling of the abdomen from an accumulation of water, and is divided into two varieties, viz. :—First, *Ascites Abdominalis*, when the water is in the cavity of the lining membrane of the abdomen, which is distinguished by the equal swelling of the abdominal walls.

Second, *Ascites Saccatus*, in which the water is in the *Ovarium*. The fluctuation in this species is less, and the swelling is not so general, particularly at first.

TREATMENT.—The principal remedies for the cure of dropsy of the abdomen are *Aconite, Nux-vom., Mercurius, Apocynum-can., Lachesis, Arsenicum, Apis-mel, Ipecac., Bryonia, Sulphur, Ferrum* and *China*, viz.:

In the commencement of this disease, or during any stage of it, when there are symptoms of fever or inflammatory action, such as an increased pulse and hot skin, give *Aconite* every three or four hours until these symptoms are removed.

When it is attended with pain and soreness in the right side, the eyes are inclined to be yellow, the skin sallow, with a disinclination to mental or physical exertion, and the bowels are constipated, *Nux-vomica* and *Mercurius* in alternation every four hours until the yellowness of the skin disappears, and the soreness of the liver is removed.

When the right side is hard and painful, the abdomen swollen and painful, as if it were bruised, and the urine scanty, *Apocynum* every three hours.

When it is attended with swelling and pain in the region of the liver, the swelling soft and tender, similar to *ramollissement* (*softening*) of the liver, the bowels hard, distended and obstinately constipated, the urine high-colored and depositing a sediment, sickness of the stomach and a disposition to vomit, *Lachesis* every four hours.

When the abdomen (*belly*) is much swollen and painful, pressing sensation in the region of the liver, the bowels constipated, or a tendency to diarrhœa, with tenesmus (a bearing down), distress and uneasiness in the pit of the stomach, a burning sensation, nausea and vomiting, *Arsenicum* as above.

Arsenicum is also especially indicated by a general swelling, sunken countenance, and great debility.

When the abdomen is much distended and there is a general dropsical effusion, the urine is passed in diminished quantities, the skin rather pale, and attended with an occasional biting and stinging sensation, *Apis-mel* three times a day.

When the disease is caused by repelled eruptions, give *Ipecac.*, *Bryonia*, and *Sulphur*, as they are here arranged; *Ipecac.* first, three doses at intervals of three or four hours, and the others in order similarly, if the first remedy proves insufficient.

When it succeeds Scarlet Fever the *Apocynumcan.*, given every two, three, or four hours, as the case required, has proved most serviceable in my practice.

When the disease is caused by a contusion or blow, *Arnica*.

When it is caused by the intemperate use of intoxicating drinks, *Lachesis* and *Nux-vomica* in alternation, or the former morning and noon, and the latter at night.

When caused by debility, *Ferrum* and *China* every six or eight hours; they may be given in alternation, or one used for a few administrations, and then the other.

DIET AND REGIMEN.—The diet should be light, unirritating, and of easy digestion, but still nutritious and of rather a solid character, because fluids ought to be avoided as much as possible. Avoid damp humid atmospheres and variable temperatures.

DROPSY OF THE BRAIN.
(*Hydrocephalus.*)

This is a disease principally confined to children, and is characterized by a great variety of symptoms during its irritative or initiatory stage. But as the disease advances, the stomach frequently becomes very irritable and is occasionally attended with retching and vomiting, the little patient is restless, and there is a constant tossing about and moaning during sleep, rolling of the head on the pillow, and a tendency to stupidity; the pupils of the eyes become dilated, and drowsiness increases, until a full coma supervenes. At this stage of the disease, palsy of one side frequently occurs, which is manifested by a tremulous motion of the arm and contraction of the fingers.

TREATMENT.—During the inflammatory stage of this disease, when the skin is hot and dry, and the pulse quick, give *Aconite.*

When the face is flushed, the eyes red, rolling of the head on the pillow, frequent startings during sleep, or waking with a scream or appearance of fright, *Belladonna.*

When effusion has taken place, the face is flushed,

DROPSY OF THE BRAIN.

the lips dry, the tongue coated with a dark yellowish fur, the bowels constipated, the abdomen swollen, the breathing hurried, and the skin hot and dry, *Bryonia.*

When there is a haggard and staring look, the eyes inclined to squinting, involuntary twitchings of the muscles, the head and body inclining backward, and an occasional groaning and crying, *Nux-vomica.*

When the body appears rigid, or there is an almost constant motion of the limbs, the patient inclines to a stupor, the pulse rather full, the face swollen, profuse involuntary discharges of urine, *Stramonium.*

When the face is pale and swollen, constant rolling of the head from side to side, the limbs rather rigid, and the pulse small and frequent, *Helleborus.*

When the face is pale, and the eyes are surrounded with a blue margin, the breathing rather difficult, with rattling of mucus in the throat, and occasional retching and vomiting, *Tart.-emet.*

When, in addition to the nausea, the pupils of the eyes are dilated and insensible, the breathing difficult and slow, *Digitalis.*

There are other remedies deserving of attention,

such as *Mercurius, Veratrum, Phosphorus* and *Chamomilla.*

Mercurius for a doughy, clay-colored face, offensive breath, apthæ or coated tongue, and an accumulation of saliva in the mouth.

Veratrum for coldness of the whole body, pale disfigured face, cold sweat on the forehead, emaciation and great debility.

Chamomilla.—When the disease is caused by teething, I have seen this remedy act like a charm, when given in alternation with *Aconite*, during the commencement of the disease.

Administration.—This disease requires the most prompt and energetic treatment. The remedies should be administered as often as every two, three, or four hours. If Tinctures are used, mix two or three drops of the medicine selected in a tumbler half full of water, stir well, and give a tea-spoonful at a dose; if powders or pillets are used, dissolve a small sized powder, or ten or twelve pillets of the first or third attenuation or potency in the same quantity of water as directed for the Tincture, and administer in the same manner.

DROPSY OF THE CHEST.
(*Hydrothorax.*)

The symptoms of this disease are not as well defined during the incipient stage as those of dropsy of the brain; but as the disease advances they become not only well marked and distressing, but alarming, such as oppression of the chest, difficult breathing, aggravated by the least exertion; when in bed, the patient is obliged to have his head and shoulders well elevated to enable him to breathe, his sleep is interrupted by frequent startings and a sense of impending suffocation; thirst is urgent, the pulse irregular, the urine scanty and high-colored, and there is swelling of the feet and legs.

TREATMENT.—When the disease arises from a congested state of the chest, or the effusion is the result of an inflammation, with cough, shortness of breath, and paroxysms of suffocation, *Aconite* every two or three hours until relief is obtained.

When there are stitches in the chest, an inability to take in a full breath, the face bloated, the patient is drowsy during the day and sleepless at night, *Bryonia* every four hours.

When there is oppressed breathing, suffocation,

or hacking cough, heaviness of the head, the face is sunken, the eyes are dim, coldness of the extremities, the mind sad and desponding, give *Lachesis* in the same manner.

When the difficulty of breathing comes on very suddenly, palpitation of the heart, aggravated by inclining the chest forward, and the symptoms are worse in the afternoon, *Spigelia* three times a day.

When the breathing is short and hurried, the face pale, the eyes dim, hiccough, nausea, pressure in the pit of the stomach, pulse small, feeble, and irregular, uneasy and unrefreshing sleep, give *Digitalis* every four hours.

When the disease is advanced, the patient emaciated, the breathing short and anxious, general coldness, when the patient complains of a burning sensation, the urine scanty and voided with pain, *Arsenicum* every four hours.

ADMINISTRATION.—The same as in other forms of Dropsy.

DROPSY OF THE JOINTS.

(Hydrops Articuli.)

This term is applied to dropsical affections and effusions within the joints.

CAUSES.—The principal causes of this affection are cold, suppressed perspiration, the abuse of mercury, and improperly treated rheumatism.

TREATMENT.—When it is caused by cold, the joints are swollen and painful, and the pains are aggravated by the least motion, *Bryonia* three times a day.

When the joints are swollen, painful and stiff, the limbs weary and restless, the pains relieved by motion and aggravated at night, and there is a general sensitiveness to cold, give *Mercurius* morning and evening.

When the swellings are principally in the knees and feet, with heat and pains of a rheumatic character, worse at night and attended with shuddering and creeping chills, give *Pulsatilla* morning and evening.

When the swelling extends from the knees to

the legs and feet, or from the wrist to the hand, and all the symptoms are aggravated in damp weather, *Lachesis* once a day or every second day.

When the swelling is confined to the knee-joint, *Sulphur* morning and evening.

When it is caused by cold from getting wet, *Pulsatilla* or *Dulcamara* morning and evening.

When caused by mercury, *Hepar-sulph.* or *Sulphur* once or twice a day.

When caused by sulphur, *Pulsatilla* once or twice a day.

ADMINISTRATION. — Administer the remedies, whether pillets, tinctures, or powders, as previously directed. I have frequently seen the greatest benefit result from sweating the part, by enveloping it in a thin sheet of india rubber, or in oiled silk. When there is much heat, a napkin applied wet in cold water will answer better.

DIET AND REGIMEN—Should not conflict with homoeopathic treatment.

DROPSY OF THE OVARIES. (*Hydrops Ovarii.*)

In this disease the effusion takes place in the internal surface of the sack or membrane, enveloping the ovary. The swelling is first observed in the right or left iliac region, unattended with pain or much constitutional disturbance; but it gradually enlarges, until it occupies nearly the entire abdomen, when serious disturbance begins, in consequence of the pressure upon the bladder, intestines and large blood vessels.

TREATMENT.—The indications in this disease are very similar to those in dropsy of the abdomen, with the exception of the symptoms which show a derangement of the liver. The principal drugs for its removal are *Mercurius, Lachesis, Digitalis, Arsenicum, Sulphur, Apocynum-can.* and *Apis-mel.* For their administration, vide Abdominal Dropsy.

DIET AND REGIMEN—As in other forms of dropsical affections.

DROPSY OF THE SCROTUM. (*Hydrocele.*)

By this term is understood a dropsical accumulation within the membranes of the scrotum : and, also, in the *tunica vaginalis*, (the membrane immediately investing the testes.) When it attacks the membranes of the scrotum, the swelling retains the impression made with the finger upon pressure. But when the *tunica vaginalis* is the seat of the disease, it has an undulating and fluctuating feeling to the touch. This disease is readily distinguished from hernia, by its transparency and fluctuations.

TREATMENT.—*Pulsatilla, Silicea* and *Sulpur*, are recommended by very respectable authority, "each remedy for eight or twelve days, every fourth day a dose." But I have recently relied almost exclusively upon *Arsenicum* and *Helleborus-niger;* give from two to three grains of the former at a dose, three times a day, and continue its use until a burning sensation is produced; then omit it and give the latter remedy, by mixing from three to five drops of the tincture in a tumbler two-thirds filled with pure cold water, well stirred, and give a large sized tea-spoonful at a dose, twice a day until the cure is completed; after which, if there is coldness of the extremities, or a want of circulation in the skin, give a few doses of *Sulphur*.

DIET AND REGIMEN—As usual in chronic diseases

DYSPEPSIA OR INDIGESTION.

Dyspepsia or indigestion depends mainly upon weakness or want of-tone of the stomach, and is characterized by a variety of symptoms: such as pain, or aching in the region of the stomach; uneasy, bloated, or distended state of the stomach after eating; sometimes nausea, flatulency, heartburn and costiveness: palpitation of the heart, depression of spirits, and langour.

TREATMENT.—The best remedies for indigestion are *Pulsatilla* and *Nux-vomica*, the former, morning and noon, and the latter at night.

In cases occurring during the summer, attended with head-ache, aversion to food, a painful distension in the pit of the stomach, and a vomiting of food; give *Bryonia* three times a day, an hour before eating.

Carbo.-veg., is a good remedy for indigestion when attended with a sour taste; tasting of the food after it has been eaten, sickness of the stomach in the morning, raising of water from the stomach at night, and an offensive diarrhœa.

Ipecac. or *Antim.* when it is attended with nausea

and vomiting, with a sense of fulness in the stomach.

If it is caused by the use of tobacco, *Nux-vomica* or *Cocculus;* if caused by excessive study, *Nux-vom., Lachesis, Pulsatilla ;* if caused by drinking cold water, *Arsenicum, Pulsatilla, Veratrum*—when caused by eating lobsters, crabs, muscles or other shell-fish, *Rhus-tox.*

ADMINISTRATION.—It is best to give the remedy about an hour before eating. When more than one remedy is affixed to a condition, give the one named first, for two or three days ; and if that does not have the desired effect, give the others in order similarly.

DIET.—Let the diet be nutritious, but of the most unirritating character.

DYSENTERY.

Dysentery is characterized by an almost constant desire to evacuate the bowels, with the voiding at each effort of a small quantity of mucus, containing little or no fecal matter. The discharges vary in appearance, are white, yellow, or green mucus, and frequently mixed with blood, and attended with the most violent bearing down and colicky pain at each effort at stool.

DYSENTERY.

TREATMENT.—At the commencement of the disease, where there is some febrile heat, and the dysenteric discharges are attended with violent bearing down pain, give *Aconite* in alternation with *Belladonna* every two or three hours.

When the discharges are mucus, some fecal matter and streaked with blood, and attended with colicky pains, *Nux-vom.* every four hours until relief is obtained.

When the pains are cutting, before and after going to stool, and the matters evacuated are the same as above, or watery and bloody, give *Merc.-cors.* every two or three hours.

When the discharges are mere white mucus, with or without chills passing over the back, *Colchic-autum* every three hours until the character of the stools become changed.

When the discharges are perfectly white mucus, and attended with a difficulty in voiding urine, *Cantharides* every two or three hours until the urinary difficulty ceases.

When the discharges are greenish yellow, watery mucus, with blood, accompanied with colicky pains, which disappear after an evacuation, *Colocynth* every three hours.

When the discharges are green, *Chamomilla* every three hours.

When the discharges are principally mucus, and the patient complains of creeping chills passing over him, *Pulsatilla* every three or four hours.

When the discharges are yellow mucus, and attended with bearing-down pain, and a general soreness, give *Staphysagria* in the same manner.

ADMINISTRATION.—Mix two or three drops of the medicine in a tumbler half full of pure cold water and give from a tea-spoonful to a table-spoonful at a dose. If pillets are used, dissolve ten or twelve in a like quantity of water and give in similar doses.

LOCAL REMEDIES.—Much relief is generally obtained from hot fomentations to the bowels, by means of cloths wrung out in hot water. And when the bearing-down is very severe and produces much suffering, make use of an occasional injection of twenty drops of *Laudanum* to one or two table-spoonfuls of starch or rice water.

DIET AND REGIMEN.—The diet should be very light and unirritating, and composed principally of soft boiled rice, arrowroot or mutton broth made very simple.

EAR-ACHE. (*Otalgia.*)

Ear-ache is generally the pain produced by an inflammation of the internal ear, and in many instances is not relieved until a suppuration takes place, with a discharge from the ear.

TREATMENT.—For ordinary ear-ache, pain of a darting or pressing character, *Pulsatilla* every two hours until it is relieved.

When it is attended with some febrile action, and the pain is of a beating, throbbing character, *Belladonna* every hour or two until the pain is mitigated.

If the pain is worse when lying in bed, and there is a disposition to perspiration, *Mercurius* every three or four hours.

If the pain is of a tearing and stinging character, and extend up toward the temple or forehead, and the suffering is greater in the morning, *Nux-vom.* three times a day.

When there is a painful aching, or a piercing, pressing pain in the ear, and pain extending to the cheek bones, *Spigelia* every hour until relieved.

If the pain is worse at night and attended with

nausea, or if caused by getting wet, *Dulcamara* every four hours.

ADMINISTRATION.—Mix two or three drops, or dissolve ten or twelve pillets, in a tumbler about one-third full of water, stir it well, and give a teaspoonful at a dose.

EXTERNAL APPLICATIONS.—The application of warmth by means of dry heat, or simple poultices, or fomentations, will generally tend to relieve the severity of the pain.

EARS, DISCHARGES FROM THE.

A running, or discharge from the ears, is a very common termination of inflammation of the internal ear, and not unfrequently succeeds scarlet fever and measles.

TREATMENT.—When the discharge is recent, and simply the termination of an acute attack of inflammation, (or ear-ache,) *Sulphur* two or three times a day will soon arrest it.

But if the discharge has continued for any length of time and becomes chronic, *Pulsatilla*, *Mercurius* and *Sulphur* should be used; one dose a day of the first remedy for five or six days; then the others similarly if necessary.

If the discharge succeeds an attack of scarlet fever, use *Belladonna* once a day for two or three days—then alternately every second day with *Hepar-sulph.*

When it follows an attack of measles, use *Pulsatilla* and *Sulphur* in alternation, every day a dose.

If the discharge is a thick, purulent matter, give *Calcarea-carb.* or *Hepar-sulphur*, as above.

If it is bloody matter, give *Pulsatilla, Mercurius* or *Lachesis*, one dose a day.

ADMINISTRATION.—Give three or four pillets at a dose, either dry on the tongue, or dissolved in a draught of water.

And keep the ears perfectly clean, by frequent washing with warm water.

EPILEPSY.

Epilepsy may be defined as chronic convulsions, with stupor, spasmodic twitching of the muscles of the face, and frothing of the mouth.

TREATMENT.—When epilepsy occurs in a full plethoric habit, restrict the diet, and give *Aconite* once a day.

When there is violent spasmodic contortion of the limbs, clenched teeth, interrupted breathing, livid face, foaming at the mouth, *Cicutæ*.

When the attack is generally during the night, with starting and convulsion of the limbs, face pale, suddenly changing to red, breathing laborious and irregular, *Belladonna*.

When the attack is more prone to return in the evening, with one side more affected than the other, the thumbs are clenched, and the patient foams at the mouth, *Hyosciamus*.

When the attack commences by the patient's crying out, and the head is drawn backwards, features distorted, and the limbs violently convulsed, *Nux-vomica*.

When it is caused by a retention, or suppression of the menses, *Sulphur* and *Pulsatilla* in alternation.

ADMINISTRATION.—Place two or three pillets on the tongue, every fifteen or twenty minutes, until the convulsion or attack is broken. But to remove a disposition to Epilepsy, give the remedy every two or three days.

DIET.—The diet should be light, unirritating and of easy digestion.

ERUPTIONS.

In prescribing or taking homœopathic remedies for the removal of eruptions, the part affected must be considered as well as the character of the eruption.

TREATMENT.—For eruptions on the face, *Causticum*, *Graphites* and *Sulphur*, given in rotation, each remedy for a week, in daily doses.

For an eruption on the lips, *Mercurius* once a day; if that does not cure in three or four days, give *Causticum* once a day.

For an eruption on the fore-arms, *Causticum* and *Staphysagria*, in alternation, every second day a dose.

When it is located on the joints, *Dulcamara*, *Lycopodium* or *Sepia* and *Sulphur* similarly.

When located on the genitals, *Mercurius*, *Rhus tox* and *Sepia*. *Mercurius* first for three or four administrations, then the other remedies in alternation, every day a dose.

When on the scrotum, *Petroleum*, *Sepia*, and *Sulphur*, in the same manner.

When the eruption appears on the scalp, *Rhus-tox* every day for a week, then *Arsenicum* similarly. If it still continues, use *Calc.-carb.* and *Sulphur*, in alternation, every day a dose.

CHARACTER OF THE ERUPTION.

For a pimply eruption *Sulphur* first, then *Rhus-tox* every day.

For a spreading eruption, *Clematis*, *Graphites*, *Merc.-sol.*, *Sulphur.*

For dry, scaly eruptions, *Arsenicum*, *Rhus-tox. Calc.-carb.*, *Allum*.

For suppurating eruptions, *Dulcamara*, *Lycopodium*, *Mercurius*, *Staphysagria*.

For a mercurial eruption, *Hepar-sulph*. once or twice a day.

For a syphilitic eruption, *Merc.*, *Nit.-acid*, *Aurum*.

ADMINISTRATION.—Give two or three pillets, or a small powder at a dose. If the medicines are in Tinctures, mix two or three drops in a tumbler one-third full of pure cold water, stir well, and give from a tea-spoonful to a table-spoonful at a dose.

When more than one remedy is named to a condition, give the first one for three or four days, one dose a day. Should that fail to produce a material improvement, give the next in order, in the same manner, and, if necessary, the next.

DIET AND REGIMEN.—The diet should be light and unirritating, and everything avoided that tends in the least to add to the grossness of the secretions.

ERYSIPELAS. (*St. Anthony's Fire.*)

Erysipelas is a febrile disease, attended with an inflammation of the skin. The *simple* variety appears in the form of an irregular stain, or blotch of a bright red color, which soon spreads to the adjoining surface, attended with burning and stinging pain, redness and some swelling.

TREATMENT.—At the commencement of the attack, and during the febrile stage, give *Aconite* and *Belladonna* in alternation, every two hours.

When the fever has abated, give *Belladonna* and *Rhus-tox* in alternation every two or three hours, until the redness and swelling subside.

If it is attended with much biting, burning, stinging and itching, and the skin has changed from a bright red to a yellowish appearance, give *Bryonia* every two or three hours.

If blisters are formed on the surface, and they appear dark, or contain a dark, dirty appearing serum, give *Arsenicum* and *Lachesis* in alternation, every two hours.

When the erysipelatous inflammation has subsided, and the skin appears rough and scaly, or scurfy, give *Sulphur* two or three times a day, until the skin begins to assume its natural appearance.

ADMINISTRATION.—Dissolve ten or twelve pillets of the medicine, or two or three drops of the tincture, in a tumbler two-thirds full of water, and give a table-spoonful at a dose to adults, and a teaspoonful to a child.

EXTERNAL APPLICATIONS.—Keep the part covered with powdered starch, or scorched flour, simply 1ᵒ shield from the air and light.

DIET.—The diet must be of the simplest kind.

FAINTING. (*Syncope.*)

Fainting is sometimes a symptom of organic disease of the heart; or it may depend upon a diminished energy of the brain from extreme debility. Those of a delicate constitution are very prone to faint from the slightest shock, or even at the sight of blood.

TREATMENT.—When fainting attacks young females of florid complexion, give *Belladonna*—if they are subject to fainting attacks, give it once a day, until the habit is corrected.

When it arises from a plethora or fulness of habit, give *Aconite* and *Belladonna* in alternation every six or eight hours.

When it is attended with palpitation of the heart, *Pulsatilla* is the remedy.

When it is caused by fright, *Opium*. By severe pain, *Veratrum*. By fear, *Ignatia*. By joy, *Coffea*.

ADMINISTRATION.—Place two or three pillets of the medicine on the tongue and allow them to dissolve, which will generally prove sufficient. But if there is a disposition to fainting from either of the causes mentioned above, continue the remedy as above directed, until the disposition to fainting passes off.

FALLING OFF OF THE HAIR. (*Alopecia.*)

Loss of the hair is in many instances hereditary; it is a common thing to see entire families lose their hair and become bald quite early in life. But it is not unfrequently caused by severe fits of sickness, the abuse of mercury, the injudicious use of *Quinine*, and frequently results from syphilis.

TREATMENT.—When it occurs after a severe fit of sickness, give *Lycopodium, Hepar-sulph., Silex.*

When it occurs during confinement from childbirth, *Calcarea, Lycopodium, Natrum-muriat., Sulph.*

When caused by grief, *Phos.-acid, Staphysagria, Lachesis, Ignatia.*

When caused by severe attacks of head-ache— *Hepar-sulph., Nit.-acid, Silex, Sepia, Sulph.*

When caused by mercury, *Hepar-sulph., Nit.-acid, Carbo.-veg.*

When caused by the abuse of quinine, *Pulsatilla, Belladonna, Hepar-sulph.*

When caused by excessive sweating, *Mercurius, Nit.-acid.*

When caused by syphilis, *Mercurius, Nit.-acid, Aurum.*

When the hair falls off from the back part of the head, *Carbo.-veg., Silex.*

From the temples, *Calc., Kali, Lycopodium.*

When it falls off in spots, *Cantharis., Phosphorus, Iodine.*

From the brows or eye lashes, *Agaricus, Belladonna, Causticum.*

For falling out of the whiskers, *Calc.-carb., Graphites, Natrum-muriat.*

ADMINISTRATION.—Administer the remedy selected once a day, or every second day, and when two or more remedies are named to a condition, give them in the order in which they are arranged, using the first named for three or four administrations; then the next similarly, and so on.

FALLING OF THE WOMB.

This is a very troublesome and not unfrequently a very distressing condition. It is characterized by a pressing and bearing down pain in the lower part of the abdomen, and a dragging pain in the small of the back, and is sometimes attended with some urinary difficulty.

TREATMENT.—The principal remedies for this

affection are *Belladonna* in the morning and *Nux vom.* at night.

ADMINISTRATION.—The same as in other uterine difficulties.

FALLING OF THE VAGINA.

This is cured by the administration of *Mercurius* once a day, or once every second day, in doses of two or three pillets at a time, or a small powder of the third trituration.

FELONS.

Felons arise from irritation between the fascia of the muscles, or the membraneous covering of the bone, producing at first a pricking sensation, similar to that caused by a brier or splinter; then heat, swelling, and the most intense beating and throbbing pain, which, if not arrested, proceeds to suppuration.

TREATMENT.—If a felon is attended with much heat and some general fever, give a few doses of *Aconite*, at intervals of an hour or two; then give *Silicea* every four or six hours until a cure is effected.

External Applications.—As soon as the heat and swelling commence, keep the part constantly enveloped in a bandage wet with cold water containing a few drops of *Arnica*, (fifteen or twenty drops to a gill of water.)

Diet—In accordance with the homœopathic rules.

FEVER. (*Febris.*)

Fever is a disease which is, (according to Dr. Hooper,) "characterized by an increase of heat, an accelerated pulse, a foul tongue, and an impaired state of several functions of the body." But it is nothing more or less than the reactive force of the vital powers against disease.

Treatment.—The only treatment necessary for the simple form of continued fever is *Aconite*, given every three or four hours.

But if the fever is attended with stitches through the head, lameness of the back, and stiffness of the muscles of the back and neck, give *Dulcamara* in alternation with the *Aconite*, as above.

If the face is red and swollen, and the mind seems wandering, give *Belladonna*, in alternation with the *Aconite*, every two or three hours until the violence of the symptoms subsides.

ADMINISTRATION.—Mix three or four drops of the medicine, or dissolve ten or twelve pillets, in a tumbler half full of pure cold water, and give from a tea-spoonful to a table-spoonful, as the condition directs.

DIET AND REGIMEN.—The diet should be light, such as thin gruel, arrow-root, and the like. Cold water only should be drunk, and the room must be kept well ventilated.

FEVER AND AGUE.

This form of fever is characterized by cold, hot, and sweating stages, succeeding each other at regular paroxysms, which are followed by an intermission.

TREATMENT.—When the thirst is equal during the cold, hot, and sweating stages; head-ache, pain in the pit of the stomach, a sudden prostration of strength, and the paroxysms occur every thirty-six hours, *Arsenicum*.

When the chills are slight, and there is a great deal of dry heat, or when the chills come on in the night, followed by thirst and vertigo, *Belladonna*.

When the cold stage predominates, and there is thirst only during the hot stage, *Bryonia*.

When the thirst is only during the cold stage, *Capsicum*.

When there is nausea and vomiting of bile, a bitter taste, and not much thirst, *Antim.-crude*.

When there is vomiting at the commencement of the cold stage, and mucous diarrhœa during the intermission, and the patient complains of soreness of the entire body, *Pulsatilla*.

When the sweating stage is not very profuse, and there is severe head-ache, or a partial blindness during the hot stage, *Natrum-mur*.

When there is no thirst during either stage, *Nitric-acid*.

When there is a partial numbness at the approach of the hot stage; tenderness and distension of the stomach and bowels, *Nux-vom*.

When the disease appears in marshy districts, and is characterized by the cold, hot, and sweating stages, *Quinine*.

ADMINISTRATION.—Mix three or four drops of the medicine in a tumbler half full of pure cold water, stir it well, and give from a tea-spoonful to a tablespoonful at a dose. If pillets are used, give five or six at a dose, every three or four hours during the intermission.

DIET AND REGIMEN.—There is no occasion for restricting the diet in this form of fever, any more than its non-interference with the medicines demands.

FEVER REMITTENT.

Remittent fever is that form which has a remission, or a very considerable diminution of the fever once or twice a day, or every second day.

TREATMENT.—When the fever is high, the skin hot and dry, with head-ache, the pulse tolerable full, give *Aconite* every three or four hours.

If there is prominent derangement of the stomach, weight in the head, with delirious talk at night, uneasiness and apprehension, *Ant.-crude*, and *Bryonia*, in alternation, every two or three hours.

If the patient complains of pain of a rheumatic character, give *Bryonia* every three hours until the pain passes off.

When there is violent heat, thirst, head-ache, or

drowsiness, and the pulse is full, give *Belladonna*, in alternation with *Aconite*, every three hours until the symptoms are mitigated.

If the fever approaches to a typhoid form, the tongue becomes dry and dark in the centre, and a disposition to low delirium ensues, give *Rhus-tox* and *Bryonia*, in alternation, every two or three hours until the symptoms become more favorable.

ADMINISTRATION.—Mix two or three drops of the medicine, or dissolve ten or twelve pillets, in a tumbler half full of pure cold water, and give from a tea-spoonful to a table-spoonful at a dose.

DIET AND REGIMEN.—The diet must be very simple, such as plain gruel, arrow-root, rice or barley water, and the patient should not be disturbed by too many questions.

FISTULA IN ANO.

The term Fistula as used in surgery is applied to a "long and sinuous ulcer that has a narrow opening, and which sometimes leads to a larger cavity, and has no disposition to heal."

The points where these abscesses burst, or open, are uncertain; sometimes in the buttock, at others near the verge of the anus, or in the perineum, and sometimes there is an external and internal opening communicating with a cavity.

TREATMENT.—When the fistulous opening is attended with a creeping sensation in the rectum, or an itching pain in the anus, use *Rhus-tox* morning and evening.

When there is an excessive itching of the anus day and night, give *Causticum* once a day.

When it is attended with an itching of the anus and a discharge of white mucus before and during a stool, or swelling about the anus of a painful character, give *Kali-carb.* once a day, or every second day a dose.

When it is attended with a discharge of matter and blood from the anus, and the sphincter muscle appears constricted, give *Lachesis* every morning or evening.

When the fistulous opening is surrounded with everted red and shining edges, use *Lycopodium* once a day.

When attended with pain and burning during the day, oozing of moisture, or a discharge of mucus from the anus, *Sepia* every evening.

If attended with violent stitches, particularly in

the evening, or a creeping sensation in the rectum as if caused by worms, *Sulphur* once a day.

When there is a constant discharge of matter, *Silicea* once a day.

When the ulcer breaks out again after having been healed, and emits a bloody lymph, instead of pus, use *Carbo.-veg.*, morning and evening for three or four days.

ADMINISTRATION.—Use five or six pillets at a dose, or when tinctures are used, mix two or three drops in a tumbler about one-third full of water, stir well, and take from a tea-spoonful to a dessert-spoonful at a dose. If powders are used, take as much as will lay on the point of a pen-knife blade at a dose.

DIET AND REGIMEN.—If the bowels are constipated (costive,) the diet ought to be rather relaxing than otherwise, but if there is a disposition to diarrhœa and debility, the food should be nutritious and bracing, but in accordance with homœopathic rules.

FLUOR ALBUS. (*Whites.*)

Leucorrhœa, or fluor albus, is a mucous, milky secretion from the vagina, which varies in consistency and color, and is sometimes attended with excoriations, smarting pain in the back and loins, debility, and frequently palpitation, and the loss of appetite.

TREATMENT.—When the discharge is copious, yellowish, and tenacious, *Aconite* morning and evening.

When it is thick like cream, and there is itching of the parts, *Pulsatilla* morning and evening.

When it is fetid, giving the linen a yellow stain, and attended with costiveness and pain in the back, *Nux-vomica* every night.

When it occurs before and after the menses; smarting and urging to urinate, *Phosphorus* night and morning.

· If the discharge is greenish, *Sepia* every morning.

When it is yellow and attended with itching and soreness, *Sulphur* once or twice a day.

When the discharge is bloody, and appears before and after menstruating, attended with flatulent and colic pains, *Cocculus* morning and evening.

When it is greenish, acrid and corrosive, *Mercurius* similarly.

ADMINISTRATION.—Mix two or three drops of the medicine, or dissolve ten or twelve of the pillets in a tumbler half full of water, and give a table-spoonful at a dose.

DIET AND REGIMEN—According to homœopathic rules.

FRACTURES AND DISLOCATIONS.

I will not lay down any general or special rules for the treatment of fractures and dislocations, for they are not only properly included within the province of the Surgeon, but it would demand a volume to treat of their varieties in detail, as they deserve.

It is well to remark, however, that a fracture or dislocation may be known by observing the extent of the injury, the distortion or position of the limb, and the inability to use it.

When either has taken place, do not attempt its reduction, but place the limb in the easiest natural position, apply cloths wet in diluted *Arnica Tinct.*, (one part of *Arnica* and three or four parts of cold water,) and send immediately for a skilful Surgeon. By all means avoid what are called natural bonesetters, who work by instinct or magic, and leave a crooked limb as a memorial of their skill and your cupidity.

FROZEN OR FROST-BITTEN.

When any part of the nose, ears, face, hands, or feet, is frozen, rub them with snow or ice for a few seconds; then immerse them in cold water, or apply cold water, gradually increasing the temperature until some color or a natural sensation returns. If violent reaction sets in, which is frequently the case, give a few doses of *Aconite* at intervals of an hour or two.

Should it produce a red, shining swelling of the part, with itching and burning, give *Agaricus* every four or six hours.

Should there be swelling, redness, burning and blistering of the part, similar in appearance to vesicular erysipelas, apply *Tinct. Cantharides* to the part.

But when the part is much swollen, and covered

with a small pustular eruption, apply *Spts. Turpentine.*

When there is not much swelling, but great pain and aching, bathe the parts frequently with *Spirits*, and give a few doses of *Rhus-tox* at intervals of two or three hours.

ADMINISTRATION.—Give five or six pillets at a dose, or if Tinctures are used, mix two or three drops in a gill of water, stir well, and give to a child a tea-spoonful at a dose; to an adult give a dessert-spoonful.

DIET AND REGIMEN.—In accordance with homœopathic restrictions when taking medicines.

GATHERED BREAST.

Gathered breast is the result of pre-established, local irritation and inflammation, which generally result from cold; and when the condition denominated *Ague in the Breast*, passes to the suppurative stage, it receives the above term.

TREATMENT.—When an inflammation of the breast is upon the point of suppuration, give *Hepar-sulph.* every hour or two.

After suppuration has taken place, use *Silicea* three times a day.

When the discharge of pus has nearly subsided, omit the *Silex*, and give *Sulphur* once or twice a day, until the breast is entirely healed.

EXTERNAL APPLICATIONS.—The numerous poultices and drawing applications, which are generally used to favor suppuration, are very prone to involve the adjacent tissue in the difficulty, and sometimes prove very destructive. If anything is necessary, it must be of the simplest character, and merely serve to protect the part from the clothing and external injury.

ADMINISTRATION.—Give a small powder at a dose, or if pillets are used, give five or six as above directed.

GLEET.

Gleet is a mere appendix to frequent attacks of, or improperly treated gonorrhœa (clap,) and frequently continues a long time after the original disease has passed off, or at least the danger of communicating it has ceased. It consists of a thin, semi-transparent mucous discharge, unattended with pain or burning.

TREATMENT.—If there is any local irritation, or tenderness of the parts during the gleety discharge, take one drop of the *Oil of Cubebs* on a piece of sugar morning and evening, and bathe the parts frequently with cold water.

But if the discharge still continues, and there is considerable weakness of the parts, use the *Ferrum-muriatis* internally, and use moderately a stringent injections.

DIET AND REGIMEN.—In accordance with treatment, but more liberal than in gonorrhœa. Take moderate exercise and avoid violent exertion, such as lifting and straining.

GOUT. (*Arthritis.*)

Gout sometimes appears suddenly, at other times it is preceded by an unusual coldness, and suppressed perspiration of the feet and legs, diminished appetite, indigestion and flatulency.

In some instances the patient is awakened from sleep by a severe pain in the first joint of the great toe. It sometimes attacks other parts of the foot such as the heel, or perhaps the entire foot.

TREATMENT.—When the attack commences with

sickness of the stomach, use *Ant.-crud.* every four or six hours.

When the attack is attended with much fever, *Aconite* every three or four hours, until the fever abates.

When it commences with shivering and creeping chills, *Pulsatilla* every four hours until the chills pass off.

When the pains are principally in the great toe, *Pulsatilla* and *Arnica* in alternation every three or four hours.

When it is in the heel, *Canabis* morning, noon and night.

When the attack commences in the foot, knee, or calf of the leg, *Bryonia* as above.

When the entire foot is equally painful, *Rhus-tox* three times a day.

When the pain and swelling is wandering, or shifts from place to place, *Pulsatilla,. Arnica* or *Nux-vomica*, as above.

When it attacks fishermen, or those exposed to getting wet, *Dulcamara* is the best remedy.

When it affects the stomach, and particularly if it occasions nausea or vomiting, use *Antimony* every half hour or hour, until the sickness of the stomach is removed. But if it is more of an intense pain, use *Nux-vom.* as above, until it is relieved.

When it passes from the feet to the heart, and is attended with fever or symptoms of inflammation, use *Aconite* every hour until the feverish symptoms subside. Then if pain continues in the region of the heart attended with palpitation and difficult breathing, use *Pulsatilla* and *Spigelia* in alternation every hour until the pain ceases.

When the kidneys become affected, use *Aconite* and *Belladonna* in alternation every hour, until the fever subsides; then use a few doses of *Cantharides* at intervals of two or three hours.

ADMINISTRATION.—Give three or four pillets at a dose, or mix two or three drops of the medicine in a tumbler half full of water, stir it well, and give from a tea-spoonful to a table-spoonful at a dose.

DIET.—The diet must be restricted, as high living is in most instances the cause of gout, particularly if assisted by stimulating drinks.

GRAVEL, (*Or Stone in the Bladder.*)

The existence of stone in the bladder is characterized by frequent inclination to urinate, with severe pain. The urine is voided drop by drop—sometimes it will start in a full natural stream, and suddenly stop by the rolling of the calculi or stone against the orifice of the neck of the bladder; and the urine is frequently colored with blood from the irritation produced by the gravelly deposit.

TREATMENT—If there is much local irritation, or any general fever, give *Aconite* every two or three hours, until it subsides. Then give the first trituration of *Kali.-carb.* in three or five grain doses, three times a day, and continue it thus, until the calculi begins to pass off in the urine. This remedy possesses a peculiar property in dissolving urinary calculi.

The attending symptoms that may arise during the treatment, such as burning and smarting in the urethra, and a degree of tension in the region of the bladder, may demand a few doses of *Phosphorus*— or severe pain in the back low down between the hips, *Nux.-vom.*

Belladonna and *Rhus-tox* may be used where there is a violent bearing down and pressing pain.

ADMINISTRATION—The remedies for the removal of the attending symptoms may be prepared by mixing two or three drops of the tincture, or ten or twelve of the pillets in a tumbler half full of water; stir well and give a dessert-spoonful at a dose every hour or two until the patient is relieved; then continue with the *Kali.-carb.* as above directed.

DIET.—In accordance with homœopathic treatment; the drinks should consist of slippery elm, flax seed, or gum arabic tea.

HEAD-ACHE.

Headache, like cough, is symptomatic of many diseases, but sometimes appears independent of other apparent affections, or at least as a predominating symptom, and as such, directs the remedy in a great measure.

TREATMENT.—For a violent head-ache, attended with fever, give *Aconite* and *Belladonna* every hour, in alternation, until it is relieved. Use the latter remedy if it is a dull, heavy, intoxicating head-ache.

When the head-ache occurs mostly in the morning, or if it is a head-ache that prevents stooping, in consequence of the pressure in the forehead, give *Bryonia* every hour or two until it is relieved.

When it is more of a lacerating pain in one side of the head, *Pulsatilla* three times a day.

For a head-ache obliging the patient to lie down, *Rhus-tox* every hour until relief is obtained.

For a severe head-ache, with sickness of the stomach; one-sided head-ache; or for a headache when the scalp feels sore to the touch, *Nux-vomica* every two or three hours.

For violent beating in the entire head, and particularly when caused by the heat of the sun, give *Lachesis* every hour until it is relieved.

For aching pain in the forehead; lacerating, throbbing head-ache, as if the eyes would be torn from their sockets, *Cocculus*.

EXTERNAL APPLICATIONS.—There is no objection to the application of cold water to the head, when it feels grateful, nor to warm or even stimulating applications to the feet, particularly when they are cold.

HEARING

The organs of hearing are delicate and complicated, and subject to many impressions which are calculated either directly or indirectly to disturb their functions and to produce deafness, partial or complete.

TREATMENT.—When hardness of hearing is attended with a tickling and roaring in the ears, use *Aconite* once a day.

When attended with humming or roaring in the ears, *Belladonna* once a day.

If attended with singing in the ears, particularly while sitting, give *Arsenicum* once or twice a day.

When the partial deafness is attended with a noise as of rushing waters, or the right ear feels closed, use *Cocculus* every second day.

For hardness of hearing, attended with a discharge from the ears, use, in alternation, one dose a day, *Belladonna* and *Pulsatilla*.

If attended with a roaring or ringing in the ears in the morning when rising, give *Nux-vomica* every night.

For partial deafness, attended with rawness and excoriations of the internal ear, use *Mercurius* once a day.

If the deafness is caused by enlarged tonsils, as is frequently the case, give *Nitric-acid* and *Iodine*. Give the former first, one dose a day for a week or two, then the latter similarly.

If caused by the suppression of cutaneous eruptions, give *Sulphur* first, then *Antimony*, *Graphites*, or *Lachesis*, as they are here arranged, one dose a day for five or six days until they have the desired effect.

ADMINISTRATION.—To an adult give five or six pillets at a dose; to a child give from two to three. But if the medicines are tinctures, mix two or three drops in a tumbler half full of pure cold water, stir it well, and give from a tea-spoonful to a table-spoonful at a dose.

DIET AND REGIMEN.—The diet must be in accordance with the condition of the patient and compatible with homœopathic treatment.

HEART-BURN. (*Waterbrash.*)

Heart-burn is one of those unpleasant symptoms which attend dyspepsia and an acid condition of the stomach; it is also a very common symptom during pregnancy.

TREATMENT.—The principal remedies in this affection are *Nux-vom.*, *Pulsatilla*, *Bryonia*, and *Carbo-veg*.

When it is attended with a bitter or sour taste, sickness of the stomach, sour eructations, raising of an acid water from the stomach, and there is an uneasy sensation in the pit of the stomach, give *Nux-vom.* morning and evening, and continue its use so long as the bowels remain constipated.

But if there is a disposition to diarrhœa, give *Pulsatilla*, as above.

When heart-burn is attended with a painful distension of the stomach, aversion to or vomiting of food, and pains in the extremities or headache, give *Bryonia* three times a day.

When there is a flatulent condition, sour taste, eructations, and tasting of the food after it has been eaten, or when attended with raising acid water from the stomach at night, use *Carbo-veg.* morning and evening.

ADMINISTRATION.—Give three or four pillets at a dose, or mix two or three drops of the medicine in a tumbler half full of cold water, stir it well, and give from a tea-spoonful to a table-spoonful at a dose.

DIET.—Some attention should be paid to diet. Avoid all kinds of greasy or oily articles of food, and also such as are indigestible.

HIP-JOINT DISEASE.

This is a scrofulous affection of the hip-joint, and during its early stages is very likely to mislead the inexperienced, in consequence of the principal pain being first experienced in the knee; and when the disease has sufficiently advanced to distort the hip, it has been mistaken for a dislocation.

TREATMENT.—This is one of those diseases which depends upon a scrofulous diathesis and must be treated by a nourishing plan, calculated to increase the solid constituents of the system. The remedies, and the diet and regimen, must be substantially the same as in other scrofulous affections of the joints. Vide White Swelling.

HOARSENESS. (*Raucedo.*)

Hoarseness is generally symptomatic of other affections. But it sometimes appears as the prominent symptom to be regarded in point of treatment, and frequently is secondary to other affections, especially to eruptive fevers and bronchial difficulties.

TREATMENT.—For acute hoarseness, with a rough dry cough, soreness, and tenacious mucus in the throat; alternate chills and flashes of heat; the patient morose and impatient, give *Nux-vomica* three times a day.

For hoarseness, with a rough voice; a burning, tickling sensation in the throat, and disposition to perspiration, *Mercury* in the same manner.

When hoarseness occurs in children, and is attended with a rough, dry cough, soreness and mucus in the throat, and with fever in the evening, give *Chamomilla* every four or six hours.

HOARSENESS.

For a hoarseness, with soreness of the throat and chest, aggravated by talking, and pains in the limbs and head, give *Bryonia* and *Rhus-tox* in alternation every two or three hours.

For hoarseness, increased by talking, or during wet weather, worse in the morning and evening, use *Carbo-veg.* or *Phosphorus* every three or four hours.

For hoarseness, with roughness and scraping in the throat, dryness of the nostrils, loss of smell or irritation of the nose, give *Sulph.* two or three times a day.

For hoarseness, continuing after measles, use *Pulsatilla*, *Carbo-veg.*, or *Sulphur*, two or three times a day.

For chronic hoarseness, associated with catarrhal symptoms, give *Phosphorus* every evening.

When the hoarseness is worse in the morning and evening; and for weakness in the organs of speech, give *Causticum* morning and evening.

For hoarseness, with a burning sensation in the throat; when the tonsils are enlarged and inflamed; and there is difficulty in swallowing, give *Belladonna* every four or six hours.

Give *Belladonna* in alternation with *Aconite* every three hours in all cases of hoarseness, associated with an inflamed condition of the throat and tonsils.

For hoarseness, secondary to Scarlatina, give *Belladonna* and *Sulphur.*

DIET AND REGIMEN.—Light and unirritating food and drinks of cold water or mucilaginous liquids, of rice, gum arabic, etc.

Use no external application, but pay a proper attention to warmth, and strictly avoid a variable temperature or a humid atmosphere.

HOOPING COUGH. (*Pertussis.*)

Hooping cough is a contagious disease, characterized by a convulsive, strangulating cough, attended with a peculiar sound, which is termed a hoop.

TREATMENT.—During the first stages of the disease, when the child appears feverish, has a dry cough, and some difficulty in breathing, give *Aconite* two or three times a day.

If the cough is severe and suffocative, especially in the evening, or worse after taking nourishment, use *Bryonia* three times a day.

When the cough is dry, and worse after midnight, accompanied with vomiting and bleeding at the nose, give *Ipecac.* and *Nux-vom.*, in alternation, every four hours.

But when the cough is loose from the commencement, with copious expectoration and frequent vomiting of mucus, use *Pulsatilla* every four hours.

When the cough is fully established, and the peculiar hoop attends each paroxysm of coughing, *Drosera* morning and evening.

ADMINISTRATION.—Dissolve six or eight pillets, or mix two or three drops of the medicine in a tumbler about one-third full of water, stir well, and give a tea-spoonful at a dose.

DIET AND REGIMEN.—The diet should be light and unirritating, but sustaining; and all exposures to variable and humid atmospheres strictly avoided.

HYDROPHOBIA. (*Canine Madness.*)

This disease is produced by the bite of a rabid animal, as a dog, cat, or fox, and persons thus bitten dread the sight of water: hence its name. There are most generally premonitory symptoms previous to the full development of madness. The wound assumes a livid appearance, its edges become raised and inflamed, and discharge a thin, watery ichor, and, if the wound has closed, the scar will become slightly elevated, burst open, and discharge a foul, thin, offensive matter, with severe pain extending throughout the entire limb or body. The patient then begins to be reserved, suspicious, gloomy, and desponding. These symptoms generally continue from two to eight days, and are followed by a perfect aversion to all liquids, especially water. The thirst is urgent, and an attempt to drink excites the most violent suffocative spasms. The secretion of saliva becomes so profuse that the patient is almost constantly spitting. As the disease advances, his expression becomes wild, furious and agonizing, and he attempts to bite every person within his reach.

TREATMENT.—*Aconite* and *Belladonna* should be

given in alternation every hour or two during the commencement of the attack; and when there is more of a trembling of the arms than violent motion; when the eyes are red and staring, with inability to swallow, give *Hyosciamus.* This remedy is more particularly indicated if the bowels are loose, and if there are involuntary discharges of urine.

Stramonium and *Lachesis* are also good remedies; the former every two hours, when the face is swollen and turgid, the eyes sparkling or swollen and protruding; the tongue swollen, with difficulty in articulating; the latter similarly, when the face is pale and sunken, the eyes inflamed; when there is an accumulation of saliva in the mouth, and an inability to swallow.

ADMINISTRATION.—In consequence of the horror for liquids, and the great difficulty in swallowing, the medicines ought to be given in the form of pillets or powders. Place five or six pillets on the tongue and allow them to dissolve, or a small powder of the medicine may be used in the same way.

HYSTERICS.

Hysterics, according to Hooper, "appears under such various shapes, imitates so many diseases, and is attended with such a variety of symptoms, that it is difficult to give a just character or definition to it."

It appears in paroxysms of laughing, crying, screaming, or in a rapid transit from one to the other, the patient sometimes violently gesticulating, biting, and pulling out the hair, and sometimes is violently convulsed, requiring three or four able-bodied men to keep her on the bed.

TREATMENT.—When this disease occurs in a young female, or in one of full, plethoric habit, give *Aconite* and *Belladonna*, in alternation, every hour until the patient is relieved.

If the patient is delicate, nervous, and subject to palpitation of the heart, give *Pulsatilla* every half hour or hour until she is relieved.

When the attack occurs in a person troubled with painful menstruation, and rather melancholy, use *Ignatia* similarly.

If there appears to be a stupefaction of the senses, give *Opium* in the same manner.

When the disease depends upon a suppression of the menses from cold, give *Pulsatilla* and *Sulphur* in alternation, every four hours.

ADMINISTRATION.—Give from three to five pillets at a dose, either dry on the tongue, or dissolved in a spoonful of water.

INFLAMMATION.

Inflammation is characterized by heat, pain, redness, attended with more or less tumefaction and fever. It is divided into two species, viz.: Phlegmonous and Erysipelatous, subdivided into acute and chronic, local and general.

Phlegmonous inflammation is known by its bright redness, tumidity and proneness to suppurate, and by its heating and pulsatory action; it has three terminations, viz.: *Resolution*, when there is a gradual abatement of symptoms; *Suppuration*, when the inflammation does not readily yield to appropriate remedies; the beating and throbbing increase, a tumor points externally, pus is formed, and rigors set in; and *Gangrene*, or mortification, when the pain abates, the pulse sinks, and cold perspiration appears.

Acute inflammation runs a rapid course; the symptoms are well defined, the pulse full and bounding, and the skin hot and dry.

Chronic inflammation is milder, and of longer duration. *Local* applies to a part or viscous; and *General* to the entire system.

Erysipelatous inflammation is of a dull red color, superficial, and merely of the skin, spreading unequally, with burning and stinging, and generally ends in vesicles or desquamation. Vide Erysipelas.

TREATMENT.—The remedies employed for the removal of inflammation are, *Aconite, Bell., Bry., Puls.,* and others; depending upon the part or viscous affected. *Aconite* is, however, the first grand remedy in the treatment of active inflammation, (the *Sthenic* of the old school.)

DIET AND REGIMEN.—The diet in all cases of inflammation must be very simple, and of a cooling character.

INFLAMMATION OF THE BLADDER.

(*Cystitis.*)

Inflammation of this organ is characterized by violent burning, lancinating, or throbbing pain in the region of the bladder, in some instances extending to the genitals and upper part of the thighs.

TREATMENT.—When attended with violent fever, give *Aconite* every two or three hours until the general fever subsides. Then, if there is constant urging to urinate, and passing of only a few drops at a time, attended with pain and burning, give *Cannabis* every two hours.

When attended with frequent ineffectual attempts to urinate, and a burning and stinging pain in the bladder, *Cantharides* every hour or two.

When the neck of the bladder is the principal seat of the inflammation, attended with a constrictive pain in the bladder and a constant urging to urinate, use *Digitalis* every two or three hours.

When attended with pain and burning, and discharges of thick turbid urine, give *Phosphorus* similarly.

When caused by getting wet, give *Dulcamara* every three hours.

ADMINISTRATION.—Dissolve five or six pillets of the medicine indicated in a tumbler one-third full of water, and give a tea-spoonful at a dose, and apply hot fomentations to the region of the bladder.

DIET.—A very light and unirritating diet, and mucilaginous drinks, such as flax-seed tea, gum arabic, or barley-water.

INFLAMMATION OF THE BOWELS.
(*Enteritis.*)

Inflammation of the bowels is known by a fixed pain, (aching or burning,) in the region of the navel, attended with fever, vomiting, and costiveness.

TREATMENT.—Give *Aconite* every two hours, until the fever abates, and a gentle moisture of the skin appears.

Belladonna every two hours, when the pains are burning, and of a cholic character; the abdomen swollen, and attended with a beating and throbbing sensation.

Lachesis every two hours, when it is attended with lacerating and cutting pains, a burning extending towards the chest, and constipated bowels.

Bryonia every two or three hours, when there is a lacerating and drawing in of the abdomen, from the hips to the stomach, followed by stitches.

Opium every hour or two, when the abdomen is distended, the bowels obstinately constipated, and the pain is of a griping character, as if physic had been taken.

ADMINISTRATION.—Dissolve ten or twelve pillets, or mix two or three drops of the medicine in a tumbler half full of cold water, stir it well, and give a table-spoonful at a dose to an adult, and a tea-spoonful to a child.

EXTERNAL APPLICATIONS.—Apply hot fomentations to the bowels.

DIET AND REGIMEN.—The diet must be light, and of a mucilaginous character—such as soft boiled rice, arrow-root, sago, or oat-meal gruel.

INFLAMMATION OF THE EAR.
(*Otitis.*)

Inflammation of the internal ear, is known by a most excruciating, throbbing pain in the ear, which frequently extends through the head, and is accompanied with fever.

TREATMENT.—When the fever is high, the pain acute, and extending through the head, give *Aconite* and *Belladonna* in alternation, every two hours.

But when the pain is of a pressing character, as if something would press out; a stinging, or a discharge of pus from the ear, *Pulsatilla* every two or three hours, until relief is obtained.

When there is pain, or soreness and excoriation of the internal ear, *Mercurius* every three or four hours.

When the acuteness of the symptoms has passed off, and there remains an itching of the outer ears, or sensitiveness to hearing, or a humming and whizzing in the ears, *Sulphur* morning, noon and night, until these symptoms are removed.

ADMINISTRATION.—Dissolve ten or twelve pillets,

or mix two or three drops of the medicine in a tumbler half full of cold water, stir it well, and give a tea-spoonful at a dose to a child, and a tablespoonful to an adult.

EXTERNAL APPLICATIONS.—Great relief is frequently obtained from the application of warmth to the part, as hot fomentations, or a simple poultice of bread and milk, applied between cloths.

INFLAMMATION OF THE EYES.
(*Opthalmia.*)

TREATMENT.—When the inflammation is of an acute character, attended with severe pain in the head, give *Aconite* every three hours.

When the eyes are very red and congested, give *Belladonna* and *Aconite* in alternation, every three hours.

When the inflammation is more of the eye-lids, with swelling, particularly of the lower lid, *Euphrasia* three times a day.

When the inflammation and pain are of a rheumatic character, give *Bryonia* and *Rhus-tox* in alternation, every two or three hours.

When the pain is of a stitching, penetrating character, extending into the head, *Spigelia*.

When it occurs in a scrofulous person, *Calc.-carb.* twice a day.

For catarrhal inflammation of the eyes, a profuse discharge of a purulent character, or agglutinating of the lids together in the morning, and swelling of the lids, *Sulphur* first, then *Phosphorus* two or three times a day.

When there is redness, sensitiveness to light, and a sensation as if a particle of sand were in the eye, *Sulphur* and *Calcarea-carb.* in alternation, every three or four hours.

ADMINISTRATION.—Dissolve five or six pillets in as many tea-spoonfuls of water, and give a tea-spoonful as directed above.

The eye should also be shielded from the air and light as much as possible.

INFLAMMATION OF THE HEART.
(*Corditis.*)

We may be thankful that this disease is not of frequent occurrence, as it is very dangerous and runs a rapid course. The symptoms are fever, pains in the region of the heart, anxious and oppressive breathing, palpitation of the heart, at times most violent and irregular; the pulse small, tense, irregular, and tremulous; difficulty in swallowing, fainting, and sudden starting in sleep.

TREATMENT.—Give *Aconite* and *Bryonia* in alternation every half hour, until there is an abatement of the general fever.

Then if the pain continues severe in the region of the heart, or there is a tensive dullness of the left half of the chest, palpitation of the heart, and difficult breathing, give *Cannabis* every hour, until the patient is relieved, or another medicine is indicated.

When the pain is more of a spasmodic character; oppression, with suffocating fits of coughing, violent palpitation, and the symptoms are aggravated when lying on the side, give *Pulsatilla* and *Nux-vom.*; the former morning and noon, and the latter at night, or every three hours during the night if there is no relief.

When there is fine stitching pain in the region of the heart, or lacerating, with constriction of the chest, violent beating of the heart, and threatened suffocation, give *Spigelia* every two hours, until relief ensues.

ADMINISTRATION.—The same as in other inflammations.

DIET AND REGIMEN.—The same as in other inflammations.

INFLAMMATION OF THE LUNGS.

(*Pneumonia.*)

Inflammation of the lungs is characterized by fever, difficult breathing, cough, a sense of weight, and pain in the chest.

TREATMENT.—The chief remedies in this disease are *Aconite* and *Bryonia* given in alternation every two or three hours, until the most prominent symptoms abate.

But when the pain is more of a stitching character, particularly if in the left side, and attended with an expectoration of mucus, streaked with blood, give *Pulsatilla* every two hours, until these symptoms are arrested.

When the cough produces much suffering, the expectoration is a bloody mucus, or a dirty muddy sputa, give *Bryonia* or *Phosphorus* in alternation with *Antimony-tart.*, every two hours.

Strictly avoid all external applications to the chest, excepting simple fomentations.

ADMINISTRATION.—Prepare the remedies as previously directed, and give them as directed above. But when the condition demanding the use of *Phosphorus* and *Antimony* exists, give the *Phosphorus* in drop doses of the third attenuation or dilution.

DIET AND REGIMEN.—The diet should be mild and unirritating; the free use of mucilaginous drinks, such as gum arabic water, slippery elm, and flaxseed tea, will be found beneficial.

INFLAMMATION OF THE KIDNEYS.
(*Nephritis.*)

This disease is denoted by fever, pain of a very acute character in the region of the kidneys, shooting along the course of the ureters, numbness of the thighs, frequent voiding of highly-colored urine, and sometimes by vomiting.

TREATMENT.—At the commencement of the disease, when the fever is high, give *Aconite* every hour or two, until the fever subsides.

Then if the pain is shooting, tearing and cutting, the emission of urine very painful, scanty, and sometimes mixed with blood, give *Cantharides* every two hours, until the symptoms are better.

But when the pain in the back has become more like a bruised or contracted feeling, and there is a painful ineffectual desire to urinate, give *Nux-vomica* every two hours.

When it is attended with a stitching, or lacerating pain in the back, and drawing, tensive pain in the spermatic cord to the testes, frequent desire to urinate, and a brick-colored or purulent sediment in the urine, give *Pulsatilla* every two hours.

When attended with a smarting and burning, before and after urinating, passing of a small quantity of urine at a time; or a discharge of mucus from the urethra, give *Cannabis* every hour, until relief is obtained.

ADMINISTRATION.—Mix two or three drops of the tincture, or dissolve ten or twelve pillets in a tumbler half full of water, stir well, and give a dessert-spoonful at a dose; if the patient is a child, give a tea-spoonful.

DIET AND REGIMEN—As in Inflammation of the Bladder.

INFLAMMATION OF THE LIVER.
(*Hepatitis.*)

Inflammation of the liver is seldom attended with such acute pain as accompanies pleurisy. It is more of a tensive, dull, and obtuse pain in the right side, and is generally attended with an inability to lie on the left.

TREATMENT.—At the commencement, if there is much fever, and pain of a stitching character, give *Aconite* every two hours until the fever abates.

INFLAMMATION OF THE LIVER.

Then, if the pains become of a pressing character, and are aggravated by motion, and the tongue is coated with a yellow fur, give *Bryonia* every three hours.

When the pain is contractive, and the region of the liver exceedingly sensitive to the touch, *Nux-vom.* every three hours.

After the general fever has been removed by *Aconite*, the disease will most commonly yield to the use of *Bryonia* in the morning, *Mercurius* at noon, and *Nux-vom.* at night. These remedies will be found to cover every indication of the disease.

But when it is caused by the continued use of *Quinine*, the remedy is *Pulsatilla;* should this fail, give *Arsenicum* every three or four hours.

ADMINISTRATION, AND DIET AND REGIMEN—The same as in other inflammatory diseases.

INFLAMMATION OF THE NOSE.

This prominent and unprotected organ is exposed to many injuries from blows, cold, etc.

TREATMENT.—For an ordinary inflammation of the nose, attended with redness, swelling, and pain, or soreness, use *Aconite* and *Belladonna*, in alternation, every two or three hours until the inflammation subsides.

But if the swelling continues and extends to the entire organ, and is attended with coryza and sneezing, use *Mercurius* every four hours.

When the swelling continues, with coryza and loss of smell, *Sulphur* three times a day.

If it is caused by mercury, as is frequently the case, use *Hepar-sulph.* three times a day.

If caused by syphilis; red or copper-colored spots appear on the nose, use *Merc.-cors.* morning and evening.

If caused by a blow or other mechanical injury, apply diluted *Tinct. Arnica* to the part.

DIET AND REGIMEN—As in ordinary local inflammation.

INFLAMMATION OF THE STOMACH.
(*Gastritris.*)

Inflammation of the stomach is characterized by fever, anxiety, heat, and pain in the region of the stomach, and an increase of all the symptoms when anything is taken into the stomach.

DIAGNOSIS.—Pain in the stomach, with a burning sensation, loathing of food, retching, vomiting, increased by taking anything warm; hiccough, pulse small and hard, tongue coated in the centre, with the edges and tip red and shining.

CAUSES.—Acrid substances, such as arsenic and corrosive sublimate, crude articles of food, unripe, indigestible fruit, drinking largely of cold water, taking ice cream, or iced fruits, when heated by exercise.

TREATMENT.—The first remedy in this disease is *Aconite*, which is to be administered every two or three hours, until there is an abatement of the fever.

Then if nausea, with occasional retching continues, and the tongue is coated white, give *Antim.-crude*, every three hours.

But if the tongue is clean, and nausea and vomiting continue, use *Ipecac.* every hour until the nausea ceases.

When the disease is caused by cold articles taken into the stomach, such as ice, iced fruits, creams, etc., use *Pulsatilla* every two hours, until it is relieved.

When it is caused by intoxicating drinks, *Nux-vomica* similarly.

ADMINISTRATION.—Place two or three pillets on the tongue, and allow them to dissolve; and follow them by a tea-spoonful of iced water.

DIET AND REGIMEN.—All food is strictly prohibited until the disease is arrested; the patient may be allowed small quantities of ice water, and even small pieces of ice if it is desired.

INFLAMMATION OF THE TONGUE.
(*Glossitis.*)

Inflammation of the tongue commences with a throbbing pain in the tongue; burning, attended with febrile symptoms, which soon become highly inflammatory. The tongue becomes hot, dry, red and swollen; the swelling increases, until it fills

INFLAMMATION OF THE TONGUE.

the entire cavity of the mouth, and frequently protrudes between the teeth.

TREATMENT.—Give *Aconite* every hour, until the high grade of inflammation and fever is subdued, or at least materially mitigated.

Then the chief reliance is upon *Belladonna* and *Mercurius*, to be given in alternation every hour or two, until the swelling begins to abate; then extend the time to every three or four hours.

But if the injudicious use of mercury has caused the disease, *Hepar-sulph.* must be used with the *Belladonna*.

If caused by a mechanical injury, wash the part frequently with diluted tincture of *Arnica*, fifteen or twenty drops to half a tumbler of water.

If caused by the bite or sting of an insect, use *Spts. of Camphor* or *Ammonia* instead of *Arnica*.

ADMINISTRATION.—The same as in other inflammations.

INFLAMMATION OF THE WOMB.

(*Metritis.*)

This disease is generally the result of cold, suppressed menstruation, or appears secondary to child-birth, and is then termed child-bed fever. It is characterized by fever and a high grade of inflammatory action, commencing with chills, headache, thirst, a quick and tense pulse, the abdomen becoming swollen, painful, and extremely tender.

TREATMENT.—Give *Aconite* and *Belladonna*, in alternation, every two hours until the fever subsides.

Then give *Belladonna* and *Bryonia* similarly until the head-ache and general nervous irritability are removed.

If the abdomen still continues distended and tender to the touch; and the bowels are constipated, give *Nux-vomica* every three hours.

But if the above condition exists, with the exception of constipation, and there is a disposition to a looseness of the bowels, use *Pulsatilla* instead of *Nux-vom.*

Should typhoid symptoms appear, *Bryonia* and *Rhus-tox* must be used, in alternation, every two or three hours until the symptoms appear more favorable.

ADMINISTRATION.—Mix two or three drops of the medicine, or dissolve ten or twelve pillets, in a tumbler half full of water, and give a table-spoonful at a dose.

EXTERNAL APPLICATIONS.—I have always used fomentations to the abdomen, and with much relief to the patient.

DIET—Must be light, and of the most unirritating character.

INFLAMMATION OF THE BRAIN.

(*Phrenitis.*)

Inflammation of the brain is characterized by high fever, violent head-ache, redness of the face and eyes, beating and throbbing of the temporal arteries, intolerance of light and sound, watchfulness and delirium.

TREATMENT.—Give *Aconite* and *Belladonna* every

two hours, in alternation, until there is a material mitigation of the symptoms.

When the fever has somewhat subsided, and there is great sensitiveness to light and sound, or if there is raving or loss of consciousness, give *Belladonna* and *Hyosciamus*, in alternation, every two hours.

When the face is red, eyes not much injected, but rather bright; the patient suddenly starting, as if frightened, *Stramonium*.

When the pains are shooting and aggravated by motion, *Bryonia* every three or four hours.

ADMINISTRATION.—Mix two or three drops of the medicine, or dissolve ten or twelve pillets, in a tumbler half full of cold water, stir it well, and give from a tea-spoonful to a table-spoonful at a dose. Apply cold applications to the head, and stimulating ones to the inferior extremities.

INFLUENZA.

Influenza is a catarrhal inflammation of the mucous membranes of the nose, frontal sinuses, and sometimes of the superior part of the air passages of the throat and lungs, producing coughing, hoarseness, sneezing, and a discharge of acrid water from the nose.

TREATMENT.—Give *Aconite* every three or four hours, for five or six administrations; if that does not cure, give *Bryonia* and *Mercurius* in alternation every four hours, which will generally effect a cure.

But if head-ache, dryness of the throat, and sneezing continue, together with a discharge from the nose, give *Phosphorus*. If there is no discharge from the nose, but a dry coryza continue, give *Causticum* every three or four hours.

ADMINISTRATION.—Prepare the remedy as previously directed, and give a dessert-spoonful to an adult, and a tea-spoonful to a child.

DIET.—Should be rather plain and unirritating.

ITCH. (*Psora.*)

Psora, or itch, is a pustulous or vesicular eruption, which appears on the wrists, thumbs, between the fingers, in the arm-pits, on the chest—and sometimes spreads to the trunk and limbs; it is communicated by contagion.

TREATMENT.—*Sulphur* is considered the specific for itch; give it three times a day, either in form of powder, tincture, or pillets.

ADMINISTRATION.—Give of the powder, as much as will lie on a three cent piece at a dose; of the pillets, five or six at a time; of the tincture, mix five or six drops in a tumbler one-third full of water, and give from a tea-spoonful to a table-spoonful at a dose.

DIET—According to homœopathic restriction.

JAUNDICE. (*Icterus.*)

This disease depends upon a derangement of the liver and biliary organs. It is characterized by yellowness of the skin and eyes, clay-colored stools, highly-colored urine, loss of appetite, general languor, disinclination to physical and mental exertion, inactivity of the bowels, and sometimes by an uneasiness in the region of the liver.

TREATMENT.—If jaundice is attended with any febrile heat, give *Aconite* every three or four hours until the increased heat passes off.

For the loss of appetite, aversion to food, pressure and uneasiness in the region of the stomach or liver, yellow skin, disinclination to mental and physical exertion, give *Pulsatilla* in the morning, *Mercurius* at noon, and *Nux-vom.* at night; and so continue them until the disease entirely disappears.

ADMINISTRATION.—In jaundice, I prepare the first triturations, given in about three grain doses, or the *Aconite, Pulsatilla* and *Nux-vom.* in tincture; mix two or three drops in a tumbler from one-third to half full of pure cold water, stir well, and give a dessert-spoonful at a dose.

DIET—Must be unirritating and of easy digestion.

LOCK-JAW. (*Trismus.*)

Lock-jaw is a spasmodic rigidity of the muscles of the jaws. It is generally preceded by the following symptoms: an uneasy sensation in the chest; slight spasmodic twitching of the muscles of the throat; difficulty in swallowing; stiffness of the muscles of the neck and shoulders; the muscles of the jaws become rigid, but not sufficiently so at first to prevent the patient from opening his mouth to some extent; the contraction, however, increases until the teeth are firmly pressed against each other; severe pain in the pit of the stomach sets in, returning at intervals of five, ten, or fifteen minutes, with the most frightful spasmodic contraction of every muscle of the body, producing the most intense agony; the countenance becomes distorted, the pulse irregular and quick, respiration hurried, the voice unnatural, the eyes dim, and the "jaws immovably locked."

TREATMENT.—The first step in the treatment of lock jaw should be to inquire into the character of the cause, and to remove, as far as possible, the presence of irritating particles of grit, rust, dirt, spiculæ of bone, needles, or whatever foreign substance may be in contact with the nerves, tendons, or fasciæ; for, in many instances, the spasms will disappear upon the removal of an irritating foreign body. I have in several instances seen well developed symptoms of Tetanus pass off by removing a part of a needle from the hand, wrist, and thumb, and pieces of glass from the foot.

When it is caused by a punctured, lacerated, or contused wound, remove (as previously stated) all irritating substances, and give *Arnica*.

And keep constantly applied to the wound, pledgets of lint, wet in diluted *Tinct. of Arnica*, (one part of the tincture to three of water.)

Should the spasms increase, and tend to the *Opisthotonos* variety, (when the body is curved backward,) with frequent cramp-like pains in the pit of the stomach, constipation, and loss of appetite, *Nux-vomica*.

When the spasms are principally of the arms and trunk, and there is drowsiness between the spasms, *Cannabis-sat.*

When there is trembling of the whole body; rigid stretching of the limbs; the head drawn backwards and the eyes distorted, *Lachesis*.

When the spasms are principally in the muscles of the back; trembling of the limbs; or confirmed *Lock-jaw*, with staring eyes and foaming at the mouth, *Laurocerasus*.

When the body is curved backwards; the patient has a furious look and foams at the mouth; a constant, violent stretching, or thrusting out of his feet and legs as if kicking or stamping; the aggravation of the symptoms returns in the evening, *Hyosciamus*.

Rhus-tox is also a very excellent remedy in the *Opisthotonos* variety, when there is great languor; an inability to remain out of bed, and extreme sensitiveness to the open air.

ADMINISTRATION.—Mix three or four drops of the medicine in a tumbler half full of pure cold water, stir it well, and give a dessert-spoonful every half hour until there is a marked improvement; then extend the time to two, three or four hours, as the improvement may indicate. If pillets are used, give five or six at a dose, as directed above.

DIET AND REGIMEN—As in other spasmodic affections.

MEASLES. (*Rubeola.*)

Measles is characterized by fever, hoarseness, dry cough, sneezing, dullness, and about the third or fourth day, by an eruption of small red points or dots, which after the third day of their appearance gradually fade, and "end in a mealy desquimation."

TREATMENT.—It is seldom necessary to resort to any further treatment than a few doses of *Aconite* at intervals of three or four hours, until the fever is moderated; then *Pulsatilla*, every four hours.

But should symptoms of congestion of the brain set in, severe head-ache, the eyes injected, the face swollen, and sudden startings, use *Belladonna* every three or four hours.

Should the cough prove troublesome and produce soreness of the chest, *Bryonia*, every three or four hours.

Should the eruption suddenly disappear and sickness of the stomach ensue, give *Ipecac.* and *Bryonia*, in alternation, every three or four hours.

When the fever and eruption have passed off, give *Sulphur* once a day for two or three days.

ADMINISTRATION.—Dissolve ten or twelve pillets in a tumbler half full of water, or mix two or three drops of the tincture in a like quantity of water, stir it well, and give from a tea-spoonful to a table-spoonful at a dose.

DIET.—In accordance with homœopathic treat-

MENSES, IRREGULAR.

TREATMENT.—When the menses continue too long, if in one of a full habit and disposed to plethora, give *Aconite* and *Belladonna* in alternation, every two or three days.

If attended with pain in the small of the back, constipation and soreness of the abdomen, give *Nux-vomica* every night.

If attended with cutting pains in the abdomen and pressure in the womb, *Chamomilla* every four or six hours.

If attended with shuddering, chills, and palpitation, *Pulsatilla* once or twice a day.

For insufficient menstruation, use *Allum* twice a day. When the menses are scanty and pale, *Carbo.-veg.* three times a day.

When attended with itching, or burning soreness in the pudendum, *Cocculus* every four or six hours.

When attended with oppressive abdominal spasms, flatulence, and lameness, *Conium* every four hours until relieved.

When it is too profuse, vide Hæmorrhage from the Womb.

ADMINISTRATION.—Mix two or three drops of the medicine, or dissolve ten or twelve pillets in about a gill of water, stir it well, and give a dessertspoonful at a dose.

MENSES, RETENTION OF.

This is more properly a delay of the first appearance of the menses.

TREATMENT.—When the patient is rather pale, delicate and slender, and troubled with occasional shooting pain through the head, palpitation of the heart, and coldness of the hands and feet, use *Sulphur* in the morning and *Pulsatilla* at night.

But when the face is flushed, and the patient complains of morning head-ache, dizziness, and occasional bleeding at the nose, *Bryonia* morning and evening.

When the retention is attended with pain, shooting up from the pudendum, and if this is attended with shuddering, and the pain is relieved by warm applications, use *Sepia* one dose a day, until it has the desired effect.

ADMINISTRATION.—Dissolve ten or twelve pillets in a gill of water, and give a large sized tea-spoonful at a dose; or mix two or three drops of the tincture in a like amount of water, and give it in the same manner.

NOTE.—Pay particular attention to the feet; keep them dry and warm.

MENSES, DIFFICULT OR PAINFUL.
(*Dysmenorrhœa.*)

Painful menstruation is characterized by severe pains in the back, loins, and abdomen, a short time previous to the appearance of the menses, which sometimes continue during the entire period.

TREATMENT.—When the pains are severe, low down in the back; and the bowels are constipated, use *Nux-vomica* three times a day.

If the pains are of a severe, bearing down character, give *Belladonna* every three or four hours.

When the pains are oppressive, and of a spasmodic character, and the flow is rather copious, *Cocculus* every three or four hours, until relief ensues.

When attended with very general pains; and cramps of the legs, give *Veratrum*, every three or four hours.

When the pains are principally in the small of the back, weakness of the limbs, extreme sensitiveness to pain, anguish and restlessness, *Coffea* every three or four hours.

ADMINISTRATION.—Mix two or three drops, or dissolve ten or twelve pillets in about a gill of water, stir it well, and give a dessert-spoonful at a dose.

DIET AND REGIMEN.—In accordance with homœopathic rules, during treatment.

MENSES, SUPPRESSION OF.
(*Amenorrhœa.*)

The term Amenorrhœa, is applied to a temporary suppression of the menses, after they have been regularly established.

TREATMENT.—When a suppression of the menses takes place from cold, and is attended with head-ache, and flushed face, use *Aconite* every three or four hours, until the fever and head-ache subside.

Then use *Pulsatilla* and *Sulphur*, one in the morning, and the other in the evening, until they have the desired effect.

But if the retention produces spasms of the chest and abdomen, give *Veratrum* every three hours, until these symptoms are relieved.

If the suppression is in consequence of a general plethora; the face is flushed, roaring in the ears, and sparkling before the eyes, use *Belladonna* three times a day, until the patient is relieved.

ADMINISTRATION.—Prepare the medicines as in the other forms of menstrual difficulties.

DIET AND REGIMEN.—The diet must be in accordance with homœopathic rules, during treatment.

MENTAL DERANGEMENT.

Under this head is embraced the different forms and grades of diseases of the mind, such as, "INSANITY, which is characterized by a deranged intellect, erroneous judgment, from imaginary perceptions, or recollections."

MANIA, (*Madness,*) a raving, furious madness, and an entire perversion of the intellectual faculties.

MONOMANIA, (*Partial Derangement,*) a form of derangement in which the person is insane upon one particular subject, and perfectly sound upon all others.

DEMENTIA, (*Without mind,*) when there appears to be an entire absence of mind; which condition may supervene slowly, or suddenly, in a mind already fully developed.

AMENTIA, (*Idiocy.*) This form of derangement is generally congenital, and depends unquestionably upon physical conformation to a great extent, or it may result from severe sickness during infancy, before the mind was even developed.

IMBECILITY, (*Mental Unsoundness,*) or a degree of defective development, generally originating after birth, and in most cases less complete than the other forms mentioned; more properly speaking, it is a weakness of intellect.

TREATMENT.—When the derangement consists in an inconsolable anguish, when the patient howls in a piteous tone, is apprehensive of death, or is fitful, sad, depressed, irritable and despairing, then gay and full of hope, give *Aconite* every four or six hours.

When the patient is desponding, and apprehensive of misfortune, has fearful illusions, sees and hears strange things, or appears stupid and childish, give *Anacardium* once or twice a day.

When the patient is in a rage, tries to escape, or evinces a disposition to commit suicide, use *Arsenicum* once or twice a day.

When the patient is disgusted with life, longs for death, and is disposed to commit suicide, is low spirited and dejected, *Aurum*.

When the patient is perfectly insane, raves, and attempts violence upon himself and others, *Belladonna* every three or four hours.

When he is in a despairing mood, is lost in

melancholy reveries, and suspicious of having committed a crime, and sensitive to noise, *Cocculus.*

For a fearful, superstitious, and hypochondriacal mood, *Conium.*

For a complete loss of sense, when the patient does not know his nearest friends, rages furiously, foams at the mouth, mutters, and screams, is reproachful, and fierce, *Hyosciamus.*

When the derangement is of a melancholic character, the patient is quiet, serious, and avoids talking, *Ignatia.*

When there appears to be a stupefaction of the senses, a state of indifference, or a delirium, with frightful fancies, *Opium.*

When the patient is gloomy, melancholy and sad, anxious and apprehensive of dying, has a disposition to weep, sometimes alternating with a hysteric laugh, and especially if this occurs during pregnancy, *Pulsatilla* every four or six hours.

When the derangement is caused by a blow, *Arnica.*

When by fright or fear, *Aconite, Belladonna, Ignatia.*

When by excessive joy, *Coffea, Opium, Pulsatilla.*

When by violent anger, *Aconite, Nux-vomica Cham., Bryonia.*

When by intoxicating drinks, *Nux-vom., Opium Hyosciam.*

When by sexual abuses, *Staphysagria, Phosphorus.*

When by debility, *China* and *Ferrum.*

ADMINISTRATION.—Administer the remedies as they appear indicated; for instance if the paroxysm is severe, give the medicine selected every two or three hours, until a manifest impression is made—but in chronic cases, characterized by milder symptoms, once or twice a day, and, in many instances every second day, is sufficiently frequent to repeat the remedies.

MILK.

For a deficiency of milk, make use of a liberal and generous diet, with plenty of milk, and take *Calc.-carb.*, one dose a day for a week; if that does not have the desired effect, take *Causticum* in the same manner.

For a suppression of milk, take *Pulsatilla* two or three times a day, particularly if the suppression is attended with palpitation of the heart.

When suppression of milk is attended with vertigo, head-ache, and a degree of fulness, give *Aconite* every four or six hours.

But if it is attended with difficult breathing, and a congested state of the chest and brain, give *Belladonna* every hour until relief ensues.

If a suppression is caused by fright, give *Opium* or *Coffea* every four or six hours.

If caused by cold, take *Chamomilla* or *Bryonia* in the same manner.

DEPRAVED QUALITY OF THE MILK.

TREATMENT.—When the milk is thin, of a bluish color, and is rejected by the infant, use *Lachesis* once a day.

When the infant refuses the breast, or vomits immediately after nursing, take *Silicea* once or twice a day.

EXCESS OF MILK.

To decrease the amount of milk when the secretion is too great, give *Pulsatilla* for two or three days, one dose a day; then *Bryonia*, and lastly *Belladonna*, in like manner.

DIET AND REGIMEN.—The diet should consist of solids principally; milk, soups, cocoa, etc., must be prohibited.

MILK ERUPTION. (*Crusta Lactea.*)

This is a disease which mostly attacks some part of the face of infants at the breast. It is characterized by an eruption, of rather broad pustules, filled with a glutinous liquor, which forms yellowish-white scabs when the pustules are ruptured. It sometimes commences on the forehead or scalp and then spreads to the face; it is seldom attended with any other consequences than the itching, which makes the infant rather restless. But it generally appears among the finest and most healthy children, and thereby removes the supposition of hereditary disease.

TREATMENT.—*Calc.-carb.*, *Sulphur*, *Kali*, and *Silex* are the remedies usually indicated.

ADMINISTRATION.—Give the remedy twice a day, commencing with *Calcarea*, and continuing each drug three or four days, until the eruption is cured.

MISCARRIAGE.

The symptoms indicating miscarriage are pains in the back and loins, pains of a bearing-down character coming on at intervals, and the appearance of hæmorrhage.

TREATMENT.—The patient should not be allowed to take the least exercise, but must be placed in a recumbent posture and take *Secale-cornutum* every half hour. Should the pains be quite pressing and bearing-down, and symptoms of being unwell increase, give *Belladonna* in alternation with the first remedy every half hour or hour, until the pains subside.

If the symptoms are caused by a strain or violent exercise, give *Rhus-tox* every hour.

But when there is not much pain, but considerable hæmorrhage of bright red blood, give *Hyosciamus* every hour; this is the remedy also if convulsions are threatened.

Should miscarriage take place and hæmorrhage continue, keep the patient perfectly quiet and give *Secale-cornu.* and *Cinnamon*, in alternation, every hour until it is arrested, to the extent that it should be.

ADMINISTRATION.—Prepare and give the remedies as in the other uterine difficulties.

DIET AND REGIMEN.—The food must be light and unirritating, and the drinks cold.

MUMPS. (*Parotidea.*)

Mumps is an inflammatory swelling of the parotid gland, characterized by swelling of the cheek, at the angle of the jaw, which sometimes extends to the neck.

TREATMENT.—If there is any head-ache, or fever attending, use *Aconite* every three hours, until it subsides.

Then *Mercurius* every three or four hours. But if the patient complains of wandering pains, with occasional chills, use *Pulsatilla* in alternation, with the *Mercurius*.

Should the swelling suddenly disappear, and the patient become stupid, or comatose, use *Opium* every two hours, and apply some hot application to the cheek and jaw.

But should it affect the brain, and the patient become delirious, use *Belladonna* and *Hyosciamus* in alternation every hour, until he is relieved.

Should the swelling suddenly recede, and pass to the testicles, (which it is very prone to do when the patient has taken cold,) or to the breasts of females, give in the former case *Pulsatilla* and *Nux-vomica* in alternation, every three hours; in the latter, use *Belladonna* every two or three hours, and apply moderate warmth to the parts.

EXTERNAL APPLICATIONS.—It is desirable to keep the cheek and angle of the jaws constantly covered with warm flannel, or dry and warmed cotton batting.

DIET—In accordance with homœopathic rules.

NETTLE RASH; OR, HIVES.

This disease is characterized by hard elevations of the skin, irregular in form, white in the centre, and generally surrounded by a diffuse, pale redness, resembling very much in appearance the sting of a nettle; and attended with an intolerable burning, smarting, stinging, and itching.

TREATMENT.—I have always succeeded with *Urtica* and *Bryonia*, given in alternation, every two or three hours.

But, when it occurs at night—for large white blotches, with tormenting itching, and burning; the blotches appear in clusters, and produce considerable swelling of the parts, give *Arsenicum* every night, until it ceases to return.

ADMINISTRATION.—Dissolve eight or ten pillets, or mix two or three drops of the medicine in about a gill of water, and give to a child a tea-spoonful at a dose, to an adult a table-spoonful.

NEURALGIA.

Neuralgia, as the term implies, is a painful affection of the nerves. The pain is very sharp, piercing, and darting.

For neuralgia of the face, vide Pain in the Face.

For neuralgia of the head, use *Coffea* every half hour, when the pain is as if a nail were driven into the brain, when the patient is exceedingly restless, weeps, and feels chilly.

Ignatia every half hour, or hour, when the pain is momentarily relieved by motion, or change of position.

Pulsatilla every three or four hours, when the pains are worse, towards evening, and during rest and sitting; for a degree of chilliness, and sometimes palpitation of the heart.

When the pain is severe, the face flushed, and there is a dizziness, or dimness of vision, use *Belladonna* or *Cocculus* every half hour, until relief is obtained.

Colocynth, for most violent, excruciating, tearing, or drawing pains on one side; or severe aching, with sickness of the stomach.

Platina is also a good remedy, when the pain is on one side; or for a violent cramp-like pain at the root of the nose.

Mercurius, when the pain darts down to the teeth, neck, and left ear; when it is worse at night, and partially relieved by pressing the head with the hands.

Veratrum, for pain of a very severe character, mostly on one side, for great weakness, the patient almost fainting when rising, cold perspiration and chilliness. *Arsenicum* is also a good remedy, and should be used, if *Veratrum* does not relieve.

For neuralgia of the stomach, vide Pain in the Stomach

For neuralgia of the back, vide Pain in the Back.

For neuralgia of the hip joint.

For a pain in the hip-joint, as if it were dislocated, or a sharp, stitching pain, give *Pulsatilla* and *Colocynth* in alternation, every three or four hours.

If the pain is lancinating in the hips, thigh, and groin, give *Arsenicum* two or three times a day.

If the pain is of a tensive character, or pressure on the hips, as if it proceeded from the small of the back, use *Lycopodium* every three or four hours.

ADMINISTRATION.—Mix two or three drops of the medicine in a gill of water, stir it well, and give a dessert-spoonful at a dose; if pillets are used, give three or four at a dose.

NODES.

Nodes are hard, circumscribed tumors, which are produced by inflammation and swelling of the membrane covering the bone, and generally appear where the integument covering the bone is the thinnest; on the shin, back of the ear, and forehead.

TREATMENT.—It is all important that these hardened tumors should be removed as speedily as possible, and everything tending to a termination by suppuration, strictly avoided, that thereby disease of the bone may be prevented.

The principal remedies are *Mercurius*, *Nit.-acid*, and *Aurum*. Give the first remedy twice a day for one week; then the second similarly, and at the end of a fortnight, should there be any of the swelling remaining, use the *Aurum* similarly. While using the remedies, paint the swelling morning and evening with the *Tincture of Iodine.*

DIET.—The diet should be perfectly simple, consisting principally of vegetables.

NURSING SORE MOUTH.

Sore mouth seems to be constitutional with some females while nursing their infants, and appears to depend upon a peculiar condition of the digestive organs and the secretions of the mouth and throat. *Causticum, Mercury,* and *Sulphur,* are the only remedies which I have ever succeeded with.

ADMINISTRATION.—Give the remedies above named in order, three times a day, continuing each one a week, until they have the desired effect.

DIET AND REGIMEN.—Diet nutritious and of easy digestion; moderate exercise in the open air.

PAIN IN THE BACK.

Pain in the back likewise occurs frequently as an isolated symptom.

TREATMENT.—For a painful stiffness in the back, give *Bryonia* every three or four hours.

For a burning pain in the left side of the lumbar vertebræ, use *Aconite* similarly.

For a pain in the small of the back, as if it were bruised and lame, use *Arnica* three times a day.

For intense, cramping pain in the small of the back, *Belladonna* is the remedy.

For ordinary pain in the small of the back, give *Calcarea-carb.* two or three times a day.

For a drawing pain in the back, use *Chamomilla*.

For insufferable pain in the small of the back, aggravated by the least motion, give *China* every two or three hours.

For drawing pain in the small of the back, with stiffness of the neck, *Kali-carb.* every three or four hours.

For a bruised, sore pain in the small of the back, *Mercurius* or *Rhus-tox* every four or six hours.

For pain in the lower part of the back, between the hips, in the os sacrum, give *Nux-vom.* morning and evening.

For pain in the back, which almost prevents stooping, and extends around the body, as if it were hooped, *Lycopodium* every four or five hours.

For a bruised pain in the inferior extremity of the back, *Phosphorus* every evening.

For pain in the back, with stiffness of the back and nape of the neck, give *Sepia* twice a day.

For violent pain in the small of the back, or for a stiff back, *Silicia* morning and evening.

For drawing pain in the small of the back, *Stramonium* as above.

ADMINISTRATION.—Mix two or three drops of the medicine in a tumbler one-third full of cold water, stir it well, and give a dessert-spoonful at a dose. If pillets are used, take three or four at a time.

For pain in the limbs. Vide Rheumatism.

PAIN IN THE CHEST.

TREATMENT.—For aching pain in the chest, only relieved for a short time by bending the trunk backwards; pain in the chest, as if the sides were drawn together, give *Aconite*.

For a stitching pain in one side of the chest, *Arnica*.

For stitches in the breast bone, or fine stitches under the collar bone, extending to the shoulder, *Belladonna*.

For stitches through the chest, *Bryonia*.

For deep stitches through the chest and about the heart, *Causticum*.

For stitches from the middle of the chest to the right side, *Cham.*, *Sepia*.

For stitches through the left side, *Ignatia*, *Phosphorus*, *Lycopodium*.

For a violent, bruising pain in the chest, *Mercurius*.

For a strong painful oppression in the middle of the chest; violent, cutting, and lacerating pain, *Spigelia*.

For a sore pain through the entire chest, *Stannum*.

For a contractive oppression of the chest, *Staphysagria*.

For violent, sharp, shooting, lancinating pains, or an aching pain in the region of the breast bone, *Veratrum*.

ADMINISTRATION.—Dissolve ten or twelve pillets in a tumbler half full of pure cold water, stir well, and give a tea-spoonful every half hour or hour, if the attack is a recent one, and acute; but if it is an old chronic difficulty, once or twice a day is sufficient. If the medicine is in form of tincture, mix two or three drops as above and give it similarly.

PAIN IN THE HEAD.

(*Vide Head-ache.*)

PAIN IN THE FACE.

(*Tic Douloureux.*)

This painful affection of the face very frequently baffles the best skill; particularly in affecting as speedy a removal of pain as is generally accomplished with homœopathic remedies.

TREATMENT.—For a tensive and lacerating pain, extending to the ear and head, *Colocynth.*

When the pain is in the cheek-bones, for violent and cutting pain, extending to the eyes, and profuse secretion of tears, *Belladonna.*

When the pains in the face are sharp stitches, with redness of the cheeks, *Cocculus.*

For a severe pain in the face at night, *Conium.*

When the face is painful, and the bones of the face are sore when touched, *Hepar-sulph.*

For a tearing pain in the face, *Mercurius.*

For a tearing pain in the cheek-bone of one side of the face, and swelling of the cheek, *Nux-vomica.*

When the pains are more in the jaws, and occur during the evening, *Phosphorus.*

For a severe pain in the bones of the cheek; or lacerating pain, with pressure in the right side, *Spigelia*.

For a tearing pain in the right half of the face, *Sulph*.

For distortion of the face, *Lachesis*.

For convulsive twitches of the muscles of the face, *Ignatia*.

ADMINISTRATION.—Mix two or three drops of the medicine, if in tincture, in a tumbler half full of water, and give a tea-spoonful every fifteen or twenty minutes, until relief is obtained. The pillets, or powders, may be used similarly.

EXTERNAL APPLICATION.—Apply hot dry cotton batting to the part; and, if the appropriate remedy does not relieve, bathe the part every fifteen or twenty minutes, with diluted *Tinct. Aconite*, five drops to a table-spoonful of alcohol.

PAIN IN THE NOSE.

For a digging pulsative pain from the left side of the mouth to the root of the nose, use *Colocynth* every hour or two, until relief is obtained.

For pain, or soreness in the nose, and black sweaty pores, or perspirable matter on the nose, *Graphites* two or three times a day.

For a crampy pressure in the root of the nose, and in the cheek-bones, give *Hyosciamus* three times a day.

For pain in the nose, as if it were sore and ulcerated, *Ignatia* every three or four hours.

For pain in the nose, proceeding from the head, *Lachesis*, as above.

For pain of a corrosive character in the nostrils, *Lycopodium* three times a day.

For pain of the borders of the nostrils, as if they were sore and ulcerated, *Nux-vomica* every three or four hours.

For constriction of the nose, use *Hellebore* twice a day.

For soreness of the nose, and scurfy nostrils, *Kali-carb.* once a day.

For swelling of the entire nose, use *Mercurius* every evening.

For soreness and ulceration of the external ring of the nose, *Pulsatilla* morning and evening.

For polypus of the nose, use *Staphysag.*, *Phos.*, *Calc.-carb.*, *Silex*, and *Sepia*.

ADMINISTRATION.—Give five or six pillets at a dose; or mix two or three drops of the medicine in a tumbler one-third full of water, stir it well, and give from a tea-spoonful to a table-spoonful at a dose. For polypus, commence with the remedies in order, as named, and give a dose a day; continuing each remedy three or four days.

PAIN IN THE STOMACH.
(*Gastralgia.*)

TREATMENT.—For a hard painful pressure after a meal, use *Belladonna*, *Pulsatilla*, or *Nux-vom.*

For a pressure in the stomach, or contractive pains, *Bryonia*.

For a spasmodic pain in the stomach, with a sense of pressure, *Carbo.-veg.*

For painful distension of the stomach, or a bloatedness in the region of the stomach, *Chamomilla*.

For an intense pain in the stomach, *Lycopodium*.

For an aching, drawing pain, increased by motion, *Pulsatilla*.

For an uneasy pressure in the pit of the stomach, as if it were swollen, *Rhus-tox*.

For a cramping, burning pain in the stomach, *Sepia*.

For a violent pressure in the pit of the stomach, for a severe, sharp, cutting piercing pain, *Veratrum*.

For spasm, or cramp of the stomach, vide Cramp of the Stomach.

ADMINISTRATION.—If the attack is recent and acute, give the remedy every ten or fifteen minutes, until relief is obtained; but if the pain is of a chronic character, once or twice a day is sufficient.

DIET.—The diet must be light, and easily digestible; and all kinds of pastry, indigestible fruit, etc., avoided.

PALPITATION OF THE HEART.

This is not only a very distressing symptom, but, in many instances, somewhat alarming. It does not, however, always denote organic disease, for it is a very common symptom in the young when there is a rapid growth, and also in those of delicate constitutions and nervous excitable temperaments.

TREATMENT.—When it occurs in young, robust persons, use *Aconite* and *Pulsatilla* every ten or fifteen minutes during the attack, and in alternation once a day for a few days, until the disposition is removed.

When it occurs during old age, give *Arsenicum* and *Lachesis* similarly.

When it occurs during pregnancy, give *Nux-vomica* and *Pulsatilla* similarly.

When the attack comes on while the patient is at rest, or after sleeping, give *Lachesis* in the same manner.

If the attack is attended with stitches in the region of the heart, use *Ignatia* and *Spigelia*, in alternation, in the same manner.

When it is caused by fright, give *Opium;* when by sudden joy, *Coffea;* when by fear, or anguish, *Veratrum;* when by intoxicating drinks, *Nux-vom.* or *Lachesis.*

ADMINISTRATION.—Dissolve six or eight pillets, or mix two or three drops of the medicine in a gill of water, stir it well, and give about a dessertspoonful at a dose.

PALSY.

Palsy is an entire loss, or a diminution of voluntary motion, sometimes accompanied with drowsiness. It usually comes on with a sudden loss of motion and sensibility of the part, but sometimes it is preceded by a degree of numbness, coldness, paleness, and slight convulsive stitches or twitches.

TREATMENT.—When this disease occurs in a full, plethoric habit, give *Aconite* two or three times, at intervals of two hours.

Should it continue, and if there is reason to suspect effusion or extravisation, give *Arnica* every two hours.

If the palsy is of one side, or of the lower limbs, with a tingling sensation of the part, give *Rhus-tox* every three hours.

When it is of the legs, and attended with trembling of the legs or sudden jerking or shocks, give *Nux-vom.* and *Cocculus*, in alternation, every three or four hours.

For paralysis of the left arm, rendering it useless,

or if attended with a numbness, coldness, and a tingling sensation, give *Aconite* every three hours.

For paralysis of the left side, or of one arm; and when the symptoms are aggravated during rest, use *Lachesis* every three hours.

For partial palsy of the arms, with pain and tingling numbness of the hands, give *Veratrum* every three or four hours.

For palsy of the tongue and organs of speech, give *Belladonna* and *Hyosciamus*, in alternation, every two or three hours. If there is not a material improvement in two or three days, use *Lachesis* instead of the *Belladonna*.

ADMINISTRATION.—The same as in other affections.

PILES. (*Hæmorrhoids.*)

External piles are small excrescences or tumors arising from a congested state of the inferior portion of the rectum, or they are situated on the verge of the anus, and are excessively painful and tender. When there are no swellings perceptible, but a degree of pain or uneasiness, and blood is passed during stool, it is termed *Blind* or *Bleeding Piles*.

TREATMENT.—For blind or bleeding piles, attended with pain in the back or colicky pain, give *Sulphur* in the morning and *Nux-vom.* at night.

When there is a flatulent state of the bowels, loss of appetite, with nausea, and the bowels are very costive, give *Antim.-crud.* morning and noon, and *Nux-vom.* at night.

When they are swollen and painful; when there is itching and burning of the anus, and if there is difficulty in urinating, use *Arsenicum* two or three times a day.

When the tumors are small, and the patient is troubled with a falling down of the rectum while at stool, and the bleeding is in fine jets, give *Acid-muriatic* every four hours.

When attended with colicky pain; when the varices are swollen and painful, with itching, and a discharge of pure blood, use *Carbo-veg.* two or three times a day.

When attended with a burning, excoriating pain in the anus and discharges of bloody mucus, give *Mercurius* three times a day.

ADMINISTRATION.—Mix two or three drops of the

medicine, or dissolve ten or twelve pillets, in a tumbler half full of water, stir well, and give a dessertspoonful at a dose.

PIMPLES ON THE FACE.

(*Acne.*)

This form of eruption generally appears on the faces of young persons at or about the age of puberty, or when the body is in full vigor; but they are sometimes caused by dissipation and other abuses.

TREATMENT.—When the eruption occurs in young people of rather delicate structure, use *Sulphur* once a day.

When it appears in the young of rather plethoric habit, fair complexion, and especially if they are disposed to phlegmonous or glandular swellings, give *Belladonna* once a day or every second day.

When it occurs in those of intemperate habits, give *Nux-vom.*, *Lachesis*, and *Ledum*, in order, one dose a day.

For the large, red Acne, use *Carbo-an.*, *Rhus-tox*, *Carbo-veg.*; and for the black Acne (*Acne punctata*) use *Belladonna, Hepar,* and *Nit.-acid,* in order, one dose a day.

When they are caused by mercury, use *Hepar-sulph.* or *Nit.-acid.*

When caused by syphilis, give *Mercury* first for a week, in daily doses; then *Nit.-acid,* similarly.

If caused by undue sexual indulgence, use *Phos.* and *Staphysagria,* in alternation, one dose a day.

ADMINISTRATION.—Give five or six pillets at a dose, or if the medicines are tinctures, mix two or three drops in a wine glass full of water, and take a tea-spoonful at a time.

PLEURISY.

Pleurisy is an inflammation of the lining membrane of the chest, and is characterized by sharp, lancinating pain piercing like a knife, inability to draw a full breath, and fever.

TREATMENT.—First give *Aconite* at intervals of an hour, for two or three times; then alternate it with *Bryonia* every three hours, until the suffering is mitigated.

After which, *Sulphur* is a very good remedy; it tends to prevent depositions in the chest, which are not uncommon in this disease.

Coffea, every night, is a very good remedy when the patient is recovering and is extremely restless at night.

When the disease is caused by an injury, give *Arnica* first, for two or three times, at intervals of an hour; then commence with the other remedies, as above directed.

ADMINISTRATION.—Mix two or three drops of the medicine, or dissolve ten or twelve pillets, in a tumbler half full of water, stir well, and give a dessert-spoonful at a dose.

EXTERNAL APPLICATIONS.—There is no objection to dry heat or hot fomentations to the side.

DIET AND REGIMEN—As in other inflammatory diseases.

PREGNANCY, DISEASES DURING.

Although pregnancy is a natural, healthy, and interesting condition, it nevertheless predisposes or gives rise to a variety of symptoms, many of which are extremely unpleasant, and sometimes very distressing.

TREATMENT.—For sickness of the stomach, aversion to food, a sweet, insipid taste, give *Ipecac.* every morning.

If the sickness of the stomach is attended with head-ache and constipation, give *Nux-vom.* morning and evening.

When the vomiting takes place after eating or drinking and is attended with weakness, give *Arsenicum* three times a day, an hour before eating.

When the sickness of the stomach is attended with loss of appetite, give *Pulsatilla* similarly.

For costiveness during pregnancy, give *Bryonia* in the morning and *Nux-vom.* at night.

For difficulty in urinating, use *Pulsatilla* morning and noon, and *Nux-vom.* at night.

For an inability to retain the urine, give *Belladonna* twice a day; if this is inefficient, give *Stramonium* in the same way; if that fails, give *Hyosciamus* and *Sepia*, in alternation, every four hours.

For sleeplessness during pregnancy give *Coffea* or *Chamomilla* every evening; and if they do not have the desired effect, use *Nux-vom.* and *Hyosciamus*, in alternation, every other night.

For fainting (hysterics), give *Ignatia* morning and noon; if it still continues, use *Pulsatilla* and *Belladonna*, in alternation, every four or six hours.

For piles during pregnancy, give *Sulphur* in the morning and *Nux-vom.* at night.

For cramps in the abdomen, use *Nux-vom.* morning and evening; should it not have the desired effect, give *Pulsatilla* and *Chamomilla*, in alternation, every four or six hours.

For convulsions during pregnancy, give *Ignatia* once or twice a day, or give it in alternation with *Hyosciamus;* should they not have the desired effect, give *Platina* once or twice a day.

For cramps in the hips, use *Colocynth* and *Rhustox*, in alternation, every four or six hours.

For cramps in the calves of the legs, use *Veratrum* twice a day, or *Secale-cornu* or *Hyosciamus*, similarly.

For cramps in the feet and toes, give *Bryonia* morning and evening, or *Lycopodium*.

For pain and swelling of the feet, use *Lycopodium* or *Bryonia* once a day.

For pain in the right side during pregnancy, use *Aconite* every four hours; should that fail, give *Chamomilla*, or give *Pulsatilla* and *Nux-vom.*, in alternation, every four hours.

For heart-burn (acid stomach or water-brash), give *Nux-vom.* every four or six hours, or *Phos.-acid*, *Pulsatilla*, or *Sulphur*, in the same manner.

ADMINISTRATION.—Mix two or three drops of the medicine, or dissolve ten or twelve pillets, in a tumbler about half full of water, and give a dessertspoonful at a dose.

DIET AND REGIMEN—In accordance with homœopathic restrictions during treatment.

SPOTS ON THE FACE DURING PREGNANCY.

It is very common for yellow or brownish spots (similar to hepatic blotches) to appear on the face during pregnancy, and sometimes to remain during the full term of gestation.. These are generally removed by *Sulphur* and *Sepia*, (the latter drug taking precedence,) a dose every two or three days, until they begin to disappear.

PRICKLY HEAT.

This is an eruptive disease, which occurs only in the summer, and consists of elevated papulæ about the size of a pin's head, of a vivid red color, and producing considerable itching; it mostly affects children and young persons.

TREATMENT.—Give two or three pillets of *Aconite* once or twice a day; if this does not relieve, give two or three pillets of *Sulphur* at night, and two or three of *Bryonia* in the morning.

QUINSY.

(Inflammation of the Tonsils.)

Inflammation of the tonsils constitutes the ordinary inflammatory sore throat, or quinsy; and is characterized by redness and swelling of the tonsils, pain and difficulty in swallowing, and sometimes of breathing.

TREATMENT.—In most instances *Aconite* and *Belladonna* given in alternation every two or three hours, will remove every symptom.

But if the tonsils continue swollen, and of a red or purple color, and attended with sharp stitches through them, or the throat is dry, with a burning sensation, alternate *Bryonia* with the *Belladonna*.

If the pain is of a stitching character, and extending to the ears when swallowing, the tongue foul and slimy, give *Mercurius* every three hours.

When small ulcers appear on the tonsils, and the pain while swallowing extends to the ears, and the throat is painful when touched, use *Ignatia* every two or three hours.

Should a disposition to suppuration appear ; and a constant ringing, with an entire inability to swallow; and liquids escape through the nostrils, give *Lahcesis* and *Hepar-sulphur* in alternation, every two hours.

ADMINISTRATION.—Mix two or three drops of the medicine, or ten or twelve pillets in a tumbler about a third full of water, stir well, and give a teaspoonful at a dose.

DIET AND REGIMEN—As in other inflammatory affections.

RED GUM.

The above term is given to a red, pimply eruption, which appears on the face, neck, and arms of infants, and sometimes spreads over the entire body. The origin of the term, or the propriety of its application to this eruption, is quite unknown to me.

The eruption is generally produced by keeping the infant too warm, and appears rather innocent in its character, requiring no other treatment than cleanliness, proper temperature, and if the infant becomes rather restless, give *Chamomilla*.

If it is exceedingly so, very sensitive, and disposed to cry, give a pillet or two of *Coffea*.

ADMINISTRATION.—Give either of the remedies, if indicated, once or twice a day.

RHEUMATISM.

Rheumatism is characterized by pains in the joints, increased by motion, swelling, redness and increased heat of the part; the pulse is generally increased in frequency, and some thirst is present during the attack. It is also quite common for the pain to pass from joint to joint, constituting the form usually denominated the *Acute Articular*. The chronic form is distinguished by pains in the joints, or muscles, without fever.

TREATMENT.—When the attack is preceded, or associated with fever, *Aconite* must be given every three or four hours, until a moderate perspiration sets in.

When the joints are red, stiff, and much swollen, attended with burning pains, give *Belladonna* every three or four hours.

When the pains are tearing, sharp, and stitching, and increased by motion, use *Bryonia*, as above.

When the pains are of a drawing character, and constant, worse at night, and the patient is very restless, give *Chamomilla* three times a day.

When the pains are jerking like shocks, or tensive, and particularly when they attack the back and loins, with a degree of numbness, and twitchings of the muscles, give *Nux-vomica*, every three or four hours.

When the pains are worse in a warm room, or when in bed, give *Pulsatilla*.

When the pains are burning and tensive; worse at night, and the part is swollen, red and shining, give *Rhus-tox*.

When the pains are lacerating and jerking, and more of the left side, and shooting from place to place, give *Colchicum*.

When it attacks a person addicted to the excessive use of ardent spirits, give *Nux-vomica*.

If caused by syphilis, give *Mercurius*.

If it is caused by getting wet, use *Dulcamara*.

If it is caused by the injudicious use of mercury, give *Hepar-sulph*.

And *Sulphur* is a very good remedy when the disease passes from the acute form to the chronic, with wandering pains from place to place.

ADMINISTRATION.—In its acute form this disease is exceedingly painful and distressing, and requires a prompt administration of remedies. They ought, therefore, to be repeated every two or three hours, until a decided improvement is perceived; then, if another remedy is not indicated by a change of symptom, give the same at longer intervals. If pillets are used, dissolve eight or ten in six table-spoonfuls of cold water, and give as above; if powders are used, prepare them similarly; but, if tinctures are used, mix from one to three drops of the remedy in a tumbler two-thirds full of pure cold water, stir it well, and give a dessert-spoonful at a dose.

EXTERNAL APPLICATIONS.—All kinds of stimulating washes, liniments, etc., must be strictly avoided. But, there is no objection to the application of silk oil-cloth, or a thin sheet of India rubber to the part, as in many instances it aids very materially by promoting local perspiration.

DIET AND REGIMEN.—The diet should be mild,

such as gruel, farina pudding, plain toast, boiled rice, simple broths, and such like; cold or damp atmospheres must be strictly avoided.

RICKETS.

Rickets is a disease most generally confined to children under four years of age, and is characterized by a large head, prominent forehead, protruding breast bone, ribs flattened on the sides, a large tumid abdomen, thin emaciated limbs, and great debility.

The muscles become relaxed, the limbs waste away, the joints enlarge, and the spinal column becomes variously distorted, producing in many instances great and most serious deformity.

TREATMENT.—This disease requires two classes of remedies, viz.: medicinal and mechanical. Of the former, the principal are *Sulphur*, *Calc.-carb.*, *Calc.-phos.*, and *Ferrum*. Give these remedies in from three to five grain doses of the first triturations, and as they are here arranged; one dose a day for a week, of the *Sulphur*, then the next in order, and so continue, until they have all been used similarly to the first.

The *mechanical means* must be constructed and adjusted according to the peculiarity of the distortion. The position should be recumbent, so as to take the weight of the body from the weakened spinal column and limbs. The patient should lie on a hard, unyielding bed, which will not accommodate itself to the form of the body.

DIET AND REGIMEN.—Generous diet, frequent cold sponging, and occasional friction with a flesh brush.

RING-WORM.

Ring-worm is a variety of Herpes (*Herpes-circinatus,*) generally met with in children, and is recognized by its annular appearance. It consists in very minute vesicles closely connected together, and forming circles varying from a quarter of an inch to two inches in diameter.

TREATMENT.—Give *Sulphur* once a day for three or four days; then, if there is any burning and itching, give *Arsenicum* similarly.

Should they still continue, and particularly if they are located on the face, use *Causticum* once a day for three or four days, and follow it by *Graphites* and *Sulphur*, in alternation, every second day a dose.

ADMINISTRATION.—To an adult give two or three pillets at a dose. If the patient is a child, dissolve the pillets in a wine glass full of water and give a tea-spoonful at a dose.

DIET—In accordance with homœopathic restrictions.

SALT RHEUM.

Salt Rheum is a popular term applied to the different varieties of Herpetic affections and Tetters, and I may say to most of the scaly eruptions unattended with fever.

TREATMENT.—For an eruption of small "red pustules, changing to ichorous, crusty, and spreading ulcers," use *Arsenicum* morning and evening.

When the eruption consists in vesicles that burst

open and form scales, which scale off, with itching, burning and stinging, give *Bryonia* three times a day.

When the eruption consists in small vesicles, surrounded by a red areola (circle), or in pimples filled with lymph appearing in clusters, and attended with a biting, stinging sensation, give *Calc.-carb.* and *Causticum*, in alternation, every six or eight hours.

When the eruption is dry, or consists of pimples filled with acrid water, and more especially if the eruption is on the arms, give *Graphites*.

When the eruption is humid, suppurating, and forming crusts, which crack open and discharge a thin ichor, use *Lycopodium*.

When the eruption is prone to spread the vesicles form and fill with a corrosive ichor; or run into each other, forming dry, scaly scabs; or crack and discharge a thin watery fluid, give *Mercurius* every morning and evening.

When the eruption runs together, and is attended with burning and itching, or for a dry, scaly eruption similar to the poison from the poison ivy, use *Rhus-tox* once or twice a day.

SALT RHEUM.

When the eruption is in the arm-pit, give *Carbo-animalis* and *Sepia*, in alternation, one dose a day.

When it appears on the face, give *Causticum*, *Graphites*, and *Sulphur*, in rotation; one dose a day of the first remedy for a week, then the next similarly, and so on until the desired effect is obtained.

When the eruption appears on the fore arms use *Graphites*, *Staphysagria*, and *Sulphur*, in like manner.

When it appears on the joints, give *Dulcamara*, one dose a day for a week, then *Lycopodium*, *Sepia*, and *Sulphur*, in the same manner.

When it attacks the genitals, give *Mercurius*, *Rhus-tox*, and *Sepia*, as above.

When it appears on the scrotum, use *Petroleum* and *Sepia*, in alternation, every six or eight hours.

For a dry and scaly eruption, use *Arsenicum*. *Allum* is also a very good remedy.

ADMINISTRATION.—Dissolve ten or twelve pillets in a gill of water, stir well, and take a dessert-spoonful at a dose. If powders or tincture are used, give them as directed in other affections.

NOTE.—A little sweet cream, or the moderate use of simple cerate to the part, when the eruption is crusty and prone to crack, will frequently prove serviceable.

SEA SICKNESS.

Sea sickness, in most instances, can be prevented by paying a little attention to diet, and taking the following remedies as they are severally indicated.

TREATMENT.—For sickness of the stomach and dizziness; when the patient feels better on lying down, give *Cocculus* every hour.

But if the patient feels better in the open air, use *Pulsatilla* every hour or two.

Nux-vomica is a very good remedy when the patient feels better in the cabin and is not exposed to the air: give it every two or three hours.

If there is almost constant vomiting, without the dizziness, use *Ipecac.* every half hour or hour, until it ceases.

But if there is constant retching and great weakness, give *Arsenicum* every hour or two.

When it is attended with costiveness, give *Nux-vomica* and *Lachesis*, in alternation, every three or four hours.

ADMINISTRATION.—Mix three or four drops of the medicine, or ten or twelve pillets in a tumbler one-third full of cold water, and give a tea-spoonful at a dose.

DIET.—The diet must be rather solid and pretty well salted, as it agrees best with the stomach in such cases. Avoid greasy or fat meats, and make use of iced water as a drink.

SEMINAL WEAKNESS.

This is a condition of the sexual system which is caused by excessive sexual indulgence, or the secret practice of self-pollution, a vice that the young are extensively addicted to, and often in consequence of a want of the proper knowledge which should be imparted to them by their parents or guardians. Its effects are hazardous in the extreme, as it lays the foundation of much physical and mental suffering.

TREATMENT.—When the patient is pale, weak, apprehensive and melancholy; is almost incapable of thinking; has pain in the back part of the head; or pains in the small of the back and weakness of the limbs, give *Nux-vom.* every night before retiring.

When in addition to the above symptoms, the patient is dull and heavy during the day and sleepless at night, use *Phosphorus* morning and evening.

If there is heaviness of the head; a desire to weep; sadness, coldness, with shuddering; loss of appetite, with distressing pressure in the pit of the stomach, give *Staphysagria* three times a day, or it may be given in alternation with *Ignatia* every six or eight hours.

When the patient is harassed at night with frightful, intimidating and raving dreams, give *Conium* every evening before retiring.

When it is attended with a degree of coldness; palpitation of the heart; gloom and melancholy; staggering gait, and trembling of the lower limbs, use *Pulsatilla* once or twice a day.

Administration—As in other affections. Sponge off the back and hips morning and evening with cold water.

Diet—Of a nutritious character, but of easy digestion, and in accordance with the remedies employed.

SCALD HEAD.

Scald Head is characterized by an eruption on the scalp, of pustules denominated *favi* and *achores*. The latter are small accumunated pustules, containing a straw-colored matter, which has the appearance and consistence of honey. The favus is larger, flatter, and contains a more viscid matter; its base is often irregular and slightly inflamed, and forms yellow scabs and scales.

Treatment.—Give *Sulphur* morning and evening for three days; then if the eruption is more of the first form. (in small pustules,) give *Rhus-tox* twice a day.

But if it is attended with much burning, and the eruption spreads and is prone to form scales and crusts, give *Arsenicum* similarly.

When it occurs to a person of decided scrofulous diathesis, give *Calcarea carb.* similarly.

ADMINISTRATION—As in other cutaneous affections.

EXTERNAL APPLICATIONS.—Keep the part washed clean with warm water and Castile soap, and apply a little sweet cream or sweet oil, so as to prevent the cracks and fissures which are frequently formed.

SCARLET FEVER.

Scarlet Fever is a contagious eruptive fever, characterized by fever, a scarlet eruption of innumerable fine red points, which become confluent and give the skin a peculiar red appearance, resembling in color the appearance of "the shell of a boiled lobster." The eruption appears first on the neck and chest, and spreads successively to the trunk and extremities; the face is swollen, the eyes are injected, and there is more or less soreness of the throat. The eruption is generally at its height on the fourth day, when the fever moderately abates and the eruption begins to fade, and on the seventh or eighth, the skin begins to desquimate or peel off, and become rough and scaly.

TREATMENT.—The superiority of homœopathic treatment above all others, in Scarlet Fever, requires no labored defence; for its success has been faithfully presented to the world by a comparison of facts. The discovery by Hahnemann, of *Belladonna*, as a specific for Scarlet Fever, places him beside the immortal Jenner, to be remembered and admired by generations yet unborn. That *Belladonna* (if properly used) is as sure a preventive of Scarlet Fever, as Vaccination is of Small Pox, there is no question. I have used it hundreds of times, and other medical gentlemen have also used it at my suggestion; yet in not a single instance have I known it to fail, and that too when I was a violent opposer of Homœopathy.

The sceptical on this point are particularly requested to use the drug during the prevalence of Scarlet Fever as an epidemic, after the following directions: mix ten drops of the tincture of the third potency in a wine glass full of water, stir it well, and to children under two years of age, give about half a tea-spoonful three times a day; to those over two years, a tea-spoonful; and to adults one drop of the tincture in a draught of cold water. Administer it thus for about two days, unless a slight vertigo, or an uneasiness about the throat is felt, or a faint rash should appear about the neck or on the point of the elbow, when it should be discontinued. At the same time observe the homœopathic diet; select your cases, give it to a part of a family, or in schools to a number that are equally subject to taking the disease, and not to the others. By so doing, if scepticism will admit of conviction, the most sceptical and prejudiced must be converted.

The medicines required in the *simple* form, are *Aconite*, *Belladonna*, and *Sulphur*.

During the initiatory or forming stage, with headache and fever, give *Aconite* every three hours until the fever is moderated.

During the eruption, give *Belladonna* every three or four hours; or if the fever is high, alternate *Aconite* with it.

After the eruption has passed off, and the skin becomes rough and scurfy, give *Sulphur* morning and evening for two or three days.

But if the disease assumes the *Anginose* variety; when there is soreness of the throat, swelling of the tonsils, much difficulty in swallowing and breathing, use *Belladonna* and *Mercurius*, in alternation, every two hours.

If the face is swollen, the mouth dry, the tongue coated with a white or yellow fur, the throat dry, with difficulty in swallowing; nausea, with or without pain in the stomach; and if the eruption loses its brightness and rather inclines to a yellow color, use *Bryonia* every three hours until the patient is better.

Or if the eruption recedes and there is nausea, give *Ipecac.* in alternation with *Bryonia*, every two or three hours.

Should there be congestion of the brain, inclining to coma; the patient sleepy and difficult to be aroused, give *Opium* every hour until the patient becomes wakeful.

During desquimation and convalescence, give *Sulphur* as previously directed.

ADMINISTRATION.—Dissolve eight or ten pillets in a wine glass full of water, stir it well, and give a tea-spoonful at a dose; or if tinctures are used, mix two or three drops in a tumbler half full of water, and give in doses similar to the above.

DIET AND REGIMEN.—The diet should be mild,

unirritating, and of a mucilaginous character; drink cold water, black tea, cocoa, crust and cracker water, and keep the apartment well ventilated and of a moderate temperature.*

SCROFULA.

Scrofula is a disease that depends upon a peculiar, degenerate condition of the system, as is very evident, from the analytical results of Dubois and other writers upon Organic and Physiological Chemistry. The disease appears in every grade of violence, from enlarged glands of the neck, armpits, groin, knee, hip-joint, to the mesenteric glands, indurated liver, tuberculated lungs, and the most loathsome ulcers.

TREATMENT.—For the removal of a scrofulous diathesis, give *Calc.-carb.*, *Kali-carb.* and *Ferrum;*

---

* Great relief is obtained by sponging the patient with water tepid, or rather inclining to cold, when the pulse is contracted and quick, the skin exceedingly hot and dry, and the mind wandering. It will serve as a powerful auxiliary in producing a gentle perspiration, relieve the cerebral oppression, and give freedom to the pulse. I have frequently practiced it, and it has always acted like a charm, producing a perfectly quiescent state, and putting the patient into a tranquil and refreshing sleep.

each remedy morning and evening, for a week, as they are here arranged, using the first triturations only.

DIET AND REGIMEN.—The diet must be of a very nutritious character; take exercise in the free open air, and avoid exposure at night, or in humid atmospheres.

Scrofulous swellings of the glands of the neck, require *Belladonna* and *Calcarea* in alternation, one dose a day for a week; then *Sulphur* once a day, until they disappear.

ADMINISTRATION.—Give two or three pillets at a dose.

For scrofulous swellings of the joints, vide Hip Joint Disease and White Swelling.

SCURVY

Scurvy is characterized by debility, a pale sickly complexion, spongy gums, which are prone to bleed; an exceedingly offensive breath, swelling of the legs, livid spots on the skin, fetid urine, and extremely offensive stools. It is a highly putrid disease, prevalent in cold climates, and on the seaboard, and especially during long voyages.

TREATMENT.—*Iodine* and *Conium* should be given in alternation, until the bleeding from the gums ceases.

Then give *Calcarea-carb.* and *Ferrum* in alternation every six hours, until the swelling around the eyes, the blanched lip, nausea and loathing of food disappear.

If the patient continues weak and debilitated, and some swelling of the limbs remains, and a dark unpleasant secretion appears on the teeth, use *Arsenicum* once a day, until he is fully restored.

ADMINISTRATION.—As in other affections.

DIET AND REGIMEN.—The diet must be nutritious

and of easy digestion; there is no objection to ripe succulent fruits. The air must be pure, and of moderate temperature.

SICKNESS OF THE STOMACH.

Sickness of the stomach, or an inclination to vomit, may arise from a variety of causes; such as a foul, irritable condition of the stomach, aversion to food from loss of appetite, debility, overloading of the stomach, irritation from worms, pregnancy and disgusting spectacles.

TREATMENT.—When sickness of the stomach is attended with tightness, pressure and fulness in the stomach, and dizziness of the head, take *Aconite* and *Belladonna* in alternation, every four or six hours.

When the sickness comes on in paroxysms, with a tendency to faintness, give *Cocculus* every three or four hours.

For sickness of the stomach, with empty eructations, and an accumulation of saliva, use *Ipecac.* once or twice a day.

For sickness in the morning, or after a meal; sour or bitter eructations (raising); pressure in the pit of the stomach, and costiveness, take *Nux-vomica*, an hour before each meal.

For sickness of the stomach upon motion, or exercise, or for nausea and vomiting immediately after drinking, take *Bryonia* once or twice a day.

For nausea and sour vomiting; painful bloatedness in the pit of the stomach, take *Chamomilla* similarly.

For continual sickness of the stomach; loss of appetite, and eructations, use *Carbo-veg.*, once or twice a day.

For nausea and vomiting, attended with severe colic pressure in the pit of the stomach, take *Cuprum* every two or three hours, until the symptoms are relieved.

For sickness of the stomach, caused by dissipation, use *Nux-vomica* or *Lachesis* every three hours; should it continue, take *Arsenicum* every four hours.

For sickness of the stomach, caused by riding in a carriage or sailing, take *Arsenicum* or *Cocculus* every three hours, or either may be taken in anticipation.

When sickness of the stomach is caused by overloading the stomach, take *Ipecac.*, *Bryonia* or *Nux-vom.*; either of them every half hour until the nausea disappears.

When it is caused by worms, give *Cina* every three or four hours.

When caused by a fall or blow, give *Arnica* every fifteen or twenty minutes until it is relieved.

ADMINISTRATION.—Mix three or four drops of the medicine, or dissolve ten or twelve pillets in a tumbler half full of water, and give to a child a teaspoonful and to an adult a table-spoonful at a dose.

DIET—In accordance with homœopathic restrictions during treatment.

SLEEP.

Want of sleep, or a general restlessness in children should be treated with *Coffea* every evening; or if there appears to be a flatulent condition of the bowels, give *Chamomilla.* *Belladonna* is also a good remedy, if there is any feverish excitement.

ADMINISTRATION.—Dissolve three or four pillets in three or four tea-spoonfuls of water, and give a tea-spoonful at a dose.

For want of sleep or restlessness in adults, give five or six pillets of *Chamomilla* at bed time. But if the restlessness is caused by sickness, or if it succeeds an attack of sickness, give *Hyosciamus*.

SMALL-POX.

Small-pox is a highly contagious eruptive disease, distinguished into two varieties, viz. the *Distinct* and *Confluent*. The former appears in distinct, elevated pustules, scattered all over the body; the latter consists of exceedingly numerous pustules, running into each other, and which are flat, irregular, and coherent.

There are four stages to be observed during the course of the disease: the febrile, the eruptive, the maturative, and that of declination or scabbing.

DIAGNOSIS.—The initiatory symptoms, before the development of the variolus fever, are very similar to those of common cold; languor, weariness, aching pains in the back and limbs, slight, creeping chills, flushes of heat, and pain in the forehead; frequently, the eyes are injected; sneezing, and a flowing coryza; thirst; sickness of the stomach, and pain in the epigastrium. These symptoms continue for a few days, when they become decidedly more inflammatory; the skin is hot and dry, the pulse full and tense, with severe pain in the head, flushed face and urgent thirst; the mind is confused and sometimes wandering, and slight hæmorrhages are apt to occur from the nose; these symptoms continue from a few hours to three or four days, when the eruption can be felt beneath the skin, like so many shot; it first appears on the forehead and about the nose and mouth, then on the arms, chest, and abdomen, and lastly, on the inferior extremities; so that in about twenty-four hours it is generally spread over the entire body. The eruption at first consists of red, elevated points, with inflamed bases, or slight areolas; toward the latter part of the second day, or the beginning of the third, the pustules begin to present central depressions, which, in a few hours, become a distinguishing mark in all of them. They gradually increase in size, and their depressed centres become more conspicuous, until the fourth day, when they appear of a disky, whitish color, surrounded by a pale, red areola; but when the pustules are very numerous or confluent, the areola or redness runs together, and gives a general appearance of redness to the interspaces. From the fifth to the seventh day, the limpid fluid which first appeared in the pustules, becomes more abundant, extends to the bases of the pustules, and gradually changes from its serous to a purulent character; at which time, the fever accompanying the eruption, becomes much less, which marks the commencement of the stage of suppuration. As this stage advances, the febrile symptoms arise, the face is frequently swollen and the eye-lids tumid, sometimes entirely closing the eyes, which occurs about the eighth day. On the tenth or eleventh day, the swelling of the face begins to recede, the hands and feet frequently swell, and the entire system becomes exceedingly sore; during this stage the patient most generally complains of soreness of the fauces and throat. About the twelfth day completes the pustule; the matter becomes more opaque and yellow, and the centre of the pustule brown, gradually extending and growing harder, until it is cemented into a brown, dry scab.

Dissication commences (like the early appearance of the eruption) on the face first; it is common to see the face covered with tolerably dry scabs, when the pustules on the extremities are hardly fully matured. When the scabs fall off they leave a red disc, which gradually disappears, leaving no traces in mild cases; but if the eruption is severe, the suppuration will frequently destroy not only the skin, but also the sub-cuticular tissue, leaving marks or pits, which disfigure the person for life.

Confluent Small Pox, is a more aggravated form; the pain in the head, back and limbs, the fever, and all the attending symptoms are necessarily more severe.

I have seen cases of this variety, in which the pustules ran into each other, so as to form a scab covering almost the entire surface; the least motion would tear them apart, producing fissures from which the blood would ooze and mix in with the scabs, producing the most loathsome appearance— the face was enormously swollen, the eyes closed, and the sufferers hardly appeared like human beings.

The eruption generally makes its appearance earlier in this variety, and is not marked with the regular course which characterizes the *Distinct* form; the matter in the pustules is dark, and sometimes acquires a corrosive character. About the eighth or ninth day of the eruption, it frequently exudes from the pustules, forming large brown scales, which fall off, from the seventh to the fifteenth day, leaving cicatrices and fissures running in every direction, and frequently producing the most hideous deformity. It is not uncommon for the fever in this variety to assume a typhoid type, running the patient down into a low state of insensibility and death. The pustules in both varieties are apt to affect the throat and eyes, which demands the closest attention.

TREATMENT.—During the first stage, when the febrile excitement is high, give a few doses of *Aconite*, at intervals of about two or three hours.

If there is severe head-ache, sensitiveness to light, and delirium, give *Belladonna* every three hours, until these symptoms pass off.

Or give *Belladonna* in alternation with *Aconite*, particularly if the above symptoms, indicating the former drug. are present in addition to those of the latter, which is most invariably the case.

When in addition to either of the above, or after the administration of *Aconite* and *Belladonna*, the patient complains of severe back-ache, and a general soreness, give *Bryonia* every three hours, until the pain is somewhat relieved.

If the patient is inclined to a stupor, is rather difficult to rouse, and sleeps with heavy breathing or snoring, give *Opium* every hour, until the patient is roused from the stupor.

During the second or eruptive stage, if the eruption does not come out free, and there appears to be a congestive state, the mind wandering, give *Stramonium* every two or three hours, until relief is obtained.

If in this stage the patient is troubled with a loose, rattling cough, and upon inspection, pustules are found forming in the throat, give *Tart.-emetic*, every three or four hours.

During the third or suppurative stage, there is frequently considerable fever, which demands a few doses of *Aconite* and *Mercurius* in alternation; the latter, more particularly, if the tongue is foul and slimy.

But should the eruption appear very dark, inclining to black, and the skin turn blue or livid; the patient's strength failing, give *Arsenicum* every two or three hours.

If in the confluent form, typhoid symptoms appear; the tongue is dark brown, sordes on the teeth, the mind wandering, great debility, give *Rhus-tox* and *Arsenicum*, in alternation every four hours.

The fourth stage (Dissication), requires an occasional dose of *Sulphur*.

The above remedies have all been used very satisfactorily in this disease. Notwithstanding, I am very much inclined to believe that Small Pox can be treated very satisfactorily with *Aconite*, *Variolin* and *Tart.-emetic* ; for, during either stage, if there is much fever, *Aconite* must constitute the chief drug, and the other two are the specifics, the one more as a prophylactic, and the other as a curative.

Aconite should be given during the initiatory symptoms, in the first stage. As soon as there is any appearance of the eruption, give the *Variolin*, two or three doses, at intervals of two or three hours, which will materially diminish the crop of pustules, and prevent a full development of those which have already made their appearance. Then give the *Tart.-emetic*, every three or four hours, until the eruption begins to scab.

In addition to the appropriate drugs in the different stages, it is frequently necessary to use some for collateral symptoms, such as extreme restlessness and inability to sleep. I have always found *Chamomilla* and *Hyosciamus* sufficient to quiet, the former, if there is any tendency to diarrhœa, and the latter in the absence of that symptom.

ADMINISTRATION.—Mix two or three drops of the medicine in a tumbler half full of pure cold water, stir it well, and give a dessert-spoonful at a dose to an adult, and a tea-spoonful to a child. If pillets are used, give five or six at a time.

The eyes should be frequently washed with cold water; and if they are closed by swelling of the lids, a small piece of linen, wet in cold water, should be kept constantly applied.

DIET AND REGIMEN.—The diet should be light, such as thin gruel, tapioca, arrow-root, and rice-water, with cooling beverages; the apartments kept clean and well ventilated.

SNUFFLES.

Snuffles or stoppage of the nose is quite common and exceedingly annoying to infants.

TREATMENT.—For dryness of the nose at night, moistness during the day, inclination to rub the nose, and for stoppage, either dry or attended with yellow matter, take *Calcarea* every three hours.

For stoppage of the nose and catarrhal symptoms, take *Sulphur* every two or three hours.

For stoppage of the nose, and dry coryza, take *Nux-vomica* every three hours.

When there is an irritation of the nose, attended with a dischârge, take *Chamomilla* every three or four hours.

For dry snuffles, worse in the open air, or when exposed to cold, take *Dulcamara* every two or three hours.

For snuffles with a discharge, particularly in the evening, or for stoppage of the left nostril, take *Carbo-veg.* every three or four hours.

ADMINISTRATION.—Dissolve three or four pillets, in a wine glass full of water, and take a small sized tea-spoonful at a dose.

SORE THROAT.

For the ordinary sore throat, produced by exposure to cold, use *Aconite* and *Belladonna*, in alternation, every two or three hours.

If there is much swelling of the tonsils, with stitches in the fauces; dryness of the throat and difficulty in swallowing, use *Belladonna* and *Bryonia* in alternation, every two hours until relieved.

If there is stitching pain in the tonsils when swallowing; pain and soreness extending to the ears, take *Mercurius* and *Ignatia*, in alternation, every two or three hours.

If there are small vesicles on the palate or tonsils; bitter taste, and complete loss of appetite, take *Sulphur* every two or three hours.

ADMINISTRATION.—Mix two or three drops of the tincture, or dissolve ten or twelve pillets, in a tumbler half full of water, and take a dessert-spoonful at a dose.

DIET—In accordance with homœopathic restrictions.

For sore throat of a more malignant character, vide Quinsy.

SORE MOUTH.

Sore mouth is frequently caused by a peculiar condition of the secretions of the stomach, and sometimes by the use of mercury. It also seems to be almost constitutional with some while nursing.

TREATMENT.—When the mouth, tongue, and fauces are red, dry, exceedingly tender, and attended with a burning sensation, use *Aconite* and *Belladonna* in alternation, every three or four hours.

When there is a white-appearing canker, give *Causticum* similarly.

When the tòngue is foul, and there is an increased flow of saliva, give *Mercurius* every two or three hours; and this is also the best remedy if there are small ash-colored cankers.

If it is caused by mercury, use *Sulphur* or *Hepar-sulph.*, three times a day.

ADMINISTRATION.—Place two or three pillets on the tongue, and allow them to dissolve, or mix two or three drops of the medicine in a tumbler one-third full of water, and take a tea-spoonful at a dose.

SORE LIPS.

TREATMENT.—When the lips are dry and cracked, use *Arsenicum* or *Veratrum*, once or twice a day.

When they are covered with small blisters, use *Ant.-tart.*, once or twice; then *Sulph.* once a day.

ADMINISTRATION.—As in Sore Mouth.

SORE NIPPLES.

It is quite common for young mothers to be severely afflicted with sore nipples. They sometimes crack and bleed, swell, become inflamed, and even threaten ulceration.

TREATMENT.—I have always succeeded better by washing the nipples immediately after nursing, with diluted *Tinct.-arnica*, five or six drops to a wine glass half full of water, (carefully washing it off with tepid water, or milk and water, before allowing the child to nurse again.) I have also found a solution of *Borax* beneficial in removing the extreme tenderness and hardening the nipple.

When the nipple is inclined to be dry, and crack, use a little sweet cream, or simple cerate, simply to lubricate and soften the part.

If there should be a disposition to scrofula, give *Calc.-carb.*, *Silicea*, and *Sulphur* in alternation, one dose a day.

ADMINISTRATION.—Give two or three pillets at a dose, or a small powder, about as much as will lie on the point of a pen-knife blade.

STRAINS, OR SPRAINS.

The first general remedy for the treatment of a sprain, or strain, is *Arnica*, particularly if it is of a limb, or joint, or the back. Bathe the part with diluted *Tinct.-arnica*, one part of the tinct. to five or six parts of water.

Should the sprain or strain, be caused by lifting a heavy weight, or by a violent exertion, give *Ruta* or *Rhus-tox* every two or three hours; or they may be given in alternation.

Should it be caused by a sudden jar, or false step, use *Bryonia* and *Rhus-tox* in alternation, every three or four hours.

ADMINISTRATION.—Mix three or four drops of the medicine in a tumbler half full of water, and give from a tea-spoonful to a table-spoonful at a dose; if pillets are used, dissolve ten or twelve in a like quantity of water, and give in the same manner.

STYES.

Styes are small tumors situated on the eyelids, generally near the greater angle, and are sometimes extremely painful.

TREATMENT.—When they are attended with much inflammation, give *Aconite* every three or four hours; then *Pulsatilla* twice a day.

If they reappear frequently, use *Calcarea-carb.* morning and evening for a week.

When suppuration has commenced, give *Silex* every four or six hours.

ADMINISTRATION.—Mix two or three drops of the medicine, in a tumbler about one-third full of water, and give from a tea-spoonful to a table-spoonful at a dose. If pillets are used, lay two or three on the tongue and allow them to dissolve.

ST. VITUS' DANCE.

This is a convulsive twitching motion of the limbs, and is generally confined to one side or limb. I have seen a case in which all the muscles appeared affected, even those of the tongue and throat.

TREATMENT.—When the disease appears in young persons of full habit, particularly in young girls, give *Aconite* and *Belladonna*, in alternation, every three hours.

If the disease does not yield in two or three days to the above treatment, and the convulsive movements are very violent, curving the body; or the limbs are violently thrown about, give *Hyosciamus* every three hours; should it still continue, use *Stramonium* similarly, or in alternation with *Hyosciamus*.

When the twitching and jerking are principally in the limbs, and continue during sleep, give *Cuprum* every four hours.

When the disease is preceded or accompanied with numbness; the movements are rather violent, with frequent distortions of the face, use *Nux-*

vomica similarly. This remedy is also more particularly indicated when the disease attacks boys.

When the disease is produced by too frequent bathing, give *Rhus-tox* three times a day.

ADMINISTRATION.—Give two or three pillets at a dose; or mix two or three drops of the tincture in a tumbler half full of water, stir well, and give from a tea-spoon to a table-spoonful at a dose.

DIET AND REGIMEN—In accordance with homœopathic restrictions.

SQUINTING. (*Strabismus.*)

Squinting is an affection of the muscles of the eye, whereby the eye is turned obliquely from the axis of vision.

TREATMENT.—When squinting exists from birth, give *Belladonna*, one dose every second or third day. Should it prove ineffectual in two or three weeks, use *Hyosciamus* in alternation with it; or use the latter drug for a week or two similarly to the first; then give *Allum* every third or fourth day, for a similar length of time.

But when squinting depends upon irritation from worms, give *Cina* two or three times, at intervals of three or four hours; should the squinting continue, give *Belladonna* and *Hyosciamus*, in alternation, every three hours.

When it occurs suddenly from palsy of the muscles of the eye, use *Rhus-tox* or *Nux-vom.* every three hours; should these remedies fail after using them two days, give *Veratrum* or *Spigelia* in the same manner.

ADMINISTRATION.—Dissolve five or six pillets, or mix two or three drops of the tincture in a gill of water, stir well, and give a tea-spoonful at a dose.

TEETHING. (*Dentition.*)

The breeding or first cutting of teeth, begins about the sixth or seventh month. The teeth appear in pairs; those of the lower jaw (the front teeth, or incisors), appear first; and the corresponding ones in the upper jaw next, and so on, until the set is completed, at about the thirteenth month; making twenty in number, and which are called the Deciduous, or Milk Teeth.

TEETHING. 261

The process of teething excites a predisposition to many diseases. Among these are irritation of the stomach and bowels, inflammation and dropsy of the brain, and convulsions.

TREATMENT.—When the teeth are tardy in cutting through the gums, give two or three pillets, or a small powder of *Calc.-carb.* once a day.

If teething excites a feverish condition, give *Aconite* every three or four hours, until the fever subsides.

If there is much heat in the head, if the gums are red and swollen, when there is sickness of the stomach, and diarrhœa, give *Belladonna;* dissolve five or six pillets in a wine glass full of water, and give a tea-spoonful every hour, until relief ensues.

If teething is attended with fever, and constipation of the bowels, give *Bryonia* morning and noon, and *Nux-vom.* at night.

If the child is threatened with convulsions, give *Belladonna, Chamomilla* or *Hyosciamus* every hour or two, until the symptoms pass off.

DIET.—Great care should be practiced in regard to diet, and everything strictly avoided that can possibly cause an irritation of the stomach and bowels.

TOOTH-ACHE. (*Odontalgia.*)

Tooth-ache is a well known disease, and makes its attacks in a variety of ways; sometimes it is very acute, with a determination to the head; at other times it is a mere soreness or gnawing pain, attended with extreme sensitiveness of the teeth and gums.

CAUSES.—There are several indirect causes. Cold and hot drinks, or food taken alternately, necessarily tend to disease the teeth. Other causes are an accumulation of tartar on the teeth, whereby the enamel is destroyed; neglect to keep them clean, the use of vegetable and mineral acids, and strong medicated dentifrices. Mercury is a prominent cause of the complete destruction of the teeth.

TREATMENT.—For throbbing pains in one side of the face, and tooth-ache, especially from cold, or congestive tooth-ache, give *Aconite.*

For violent tearing in the teeth of the lower jaw, increased by cold; violent tearing in the right side of the lower jaw, give *Agaricus.*

For dull drawing in the upper and right row of teeth all night; or for rheumatic tooth-ache, give *Belladonna*.

For tooth-ache caused by either hot or cold things, or for gnawing tooth-ache, give *Calcarea-carb*.

For tooth-ache aggravated by warm drinks; innietolrable tooth-ache, or tooth-ache worse at night, use *Cham*.

For throbbing tooth-ache, looseness of the teeth, give *China*.

For jerking tooth-ache, give *Hepar-sulph*.

For tooth-ache only when eating, give *Kali-carb*.

For a lacerating, tearing tooth-ache, give *Lachesis*.

For tooth-ache only at night, when the teeth are very painful when touched, give *Lycopodium*.

For violent tooth-ache in the night, looseness of the teeth, the gums receding from them, give *Mercurius*.

For tooth-ache occurring after dinner, lacerating tooth-ache, brought on again by cold water, give *Nux-vomica*.

For tooth-ache, with swelling of the cheek, give *Phosphorus*.

For tooth-ache aggravated by taking anything warm into the mouth, or for tooth-ache returning upon eating, give *Pulsatilla*.

For drawing tooth-ache, extending to the ear, particularly if from a hollow tooth, give *Sepia*.

For throbbing, lacerating tooth-ache, aggravated by cold water, give *Spigelia*.

For tooth-ache caused by a draft of air, or when the teeth feel elongated, give *Sulphur*.

For tooth-ache during pregnancy, give *Sepia*.

For tooth-ache relieved by cold drinks, give *Bryonia, Pulsatilla*.

For tooth-ache relieved by warm drinks, give *Lycopod., Sulph.*

For tooth-ache relieved by smoking, give *Mercury*.

For tooth-ache aggravated by smoking, give *Ignatia, Bryonia*.

For tooth-ache relieved by warmth, give *Merc., Nux-vom., Sulph.*

For the pain which is sometimes severe after a tooth has been extracted, give *Hyosciamus*.

ADMINISTRATION.—Give two or three pillets at a dose, or if the medicines are in the form of tincture, mix two or three drops in a tumbler about one-third full of pure cold water, stir it well, and give a tea-spoonful every fifteen or twenty minutes, if the attack is very severe, until it is relieved; at the same time in addition to the drug, apply heat or cold, as may be indicated. Most of the above remedies will act like a charm; if they fail, as they sometimes do, Dr. Hempel's plan may be resorted to, viz., carefully fill the hollow of the tooth with a bit of cotton, wet with one or two drops of the *Tinct.-aconite*; if this affords only temporary relief, apply to a good dentist.

DIET AND REGIMEN.—These are not of much consequence, so long as they do not interfere with the medicines. Temperature as may be most agreeable.

TYPHUS FEVER.

Typhus is a species of continued fever of low grade, with a tendency to putrefaction. It is distinguished from the inflammatory by the smallness of the pulse and great prostration from the commencement of the attack.

TREATMENT.—At the commencement, when there are alternate chills, with flashes of heat; pain in the head, back and limbs, give *Bryonia* and *Pulsatilla*, in alternation, every two hours.

Should the above remedies not arrest the disease at once, and the fever increase with the other suffering, head-ache and pain in the back and limbs, give *Aconite* and *Bryonia* in alternation.

Should the symptoms still resist treatment and become more typhoid, the skin dry, the face flushed and rather dark; the tongue dry; the pulse small and frequent, the mind wandering, use *Bryonia* and *Rhus-tox*, in alternation, every three hours.

At any time during the course of the disease, if the patient should become exceedingly drowsy and stupid, give a dose or two of *Opium*.

Should the patient become exceedingly restless or delirious, and the discharges from the bowels and urinary organs pass off involuntarily, give *Hyosciamus* every three or four hours.

ADMINISTRATION.—Mix two or three drops of the medicine or ten or twelve pillets, in a tumbler half full of water, stir well, and to an adult give a table-spoonful at a dose; to a child a tea-spoonful.

DIET AND REGIMEN.—The diet must be very light, such as thin gruel, rice or barley water; and the apartment kept well ventilated.

ULCERS.

The causes of ulcers are very various, as anything that tends to produce an inflammation of a part, contused wounds, etc., or a specific irritation of the absorbents, from scurvy, cancer, mercury, venerial or scrofulous virus.

TREATMENT.—When the ulcer is caused by the injudicious use of mercury, give *Hepar-sulphur* twice a day until it is cured.

When it is of a scrofulous character, use *Calc.-*

carb *Belladonna*, and *Ferrum*; each remedy to be given in rotation, three or four doses at intervals of twelve or twenty-four hours.

For syphilitic ulcers, use *Mercurius* twice a day for a week; then *Aurum* in the same manner, and if the cure is not completed, give *Nitric-acid* morning and evening until it is cured.

STRUCTURE, SHAPE AND APPEARANCE.—For fistulous ulcers (having a small sinus opening,) use *Lycopodium* and *Antimony*, in alternation, every six or eight hours.

For superficial ulcers, give *Lachesis*, *Mercurius*, and *Arsenicum*, in alternation, one dose a day.

For hard ulcers, with indurated or calloused edges, give *Mercurius* and *Lachesis*, in alternation, every six or eight hours.

For cancerous-appearing ulcers, give *Arsenicum* and *Conium*, in alternation, every four or six hours for a week; then *Lachesis*, one dose a day for a fortnight.

For fungous ulcers, use *Carbo-animalis* and *Phosphorus*, in alternation, every six or twelve hours.

For varicose ulcers, use *Carbo-veg.* and *Pulsatilla* in the same way.

For deep ulcers, use *Mercurius* and *Nitric-acid*, in alternation, every six hours; should these not heal them in a week or produce a marked improvement, give *Calcarea* and *Silex* in the same way.

For yellow mattery-appearing ulcers, give *Calc.-carb.* and *Silex*, in alternation every four or six hours.

For unclean, foul ulcers, use *Lachesis* once a day; should that fail, use *Kreosote* once a day.

For ulcers disposed to bleed, use *Carbo-veg.* and *Conium* in alternation every six hours, or *Phosphorus* once a day.

For ulcers, with burning, stinging pains use *Arsenicum* and *Rhus-tox*, in alternation, every six or eight hours, or *Carbo-veg.* and *Mezerium* in the same manner.

ADMINISTRATION.—Mix two or three drops of the medicine, in a tumbler half full of water, stir it well, and give from a tea-spoonful to a table-spoonful at a dose. If the medicine is in pillets, take two or three of them at a time; if in powder, take a small powder.

EXTERNAL APPLICATIONS — Simply keep the

part well cleansed by frequent washings with warm water and Castile soap, and dress with simple dressings.

DIET AND REGIMEN—In accordance with homœopathic rules.

URINE, IMMODERATE FLOW OF.
(*Diabetes.*)

TREATMENT.—When the emission of urine is copious and watery, give *Aurum* three times a day.

When the emission is in small quantities, but very frequent, and the urine of a light straw-color, give *Asparagus* similarly.

When there are copious and frequent discharges of watery urine, alternating with a diminished flow, use *Digitalis* similarly.

When the discharges are frequent and copious, attended with a dull pain in the bladder, give *Lachesis* morning and evening.

ADMINISTRATION.—Mix two or three drops of the medicine in a tumbler half full of water, and give a table-spoonful at a dose to an adult, and a teaspoonful to a child. If pillets are used, dissolve ten or twelve in a similar quantity of water, and give similarly.

Diet—In accordance with homœopathic rules.

ENURESIS NOCTURNL

(*Wetting the Bed.*)

To correct this troublesome and annoying practice, give either *Silex, Sepia, Sulphur* or *Carbo-veg.*; one dose every night, just before retiring. If the first should fail of success, after three or four administrations, give the next, and so on, until the desired effect is obtained.

SUPPRESSION OF URINE.

(*Anury.*)

Treatment.—When there is a partial or complete suppression of urine, and attended with a feverish condition, give *Aconite* every two or three hours, until the fever abates. If a severe bearing down and pressing pain accompany the above, give *Belladonna* in alternation with the *Aconite.*

When there is a frequent and almost constant urging, and passing of a very small quantity at a time of high-colored urine, give *Arsenicum* every three hours.

When it is attended with the most violent burning and cutting pains, and the urine is voided in drops, and sometimes streaked with blood, give *Cantharis* every hour, until relieved.

When the difficulty is attended with an itching in the urethra, a painful desire to urinate; and especially if it is in consequence of spasmodic stricture or irritation from piles, *Nux-vomica* every hour or two, until it is relieved.

Note.—I have never had the *Apocynum-cannabinum* fail, in partial or complete retention of urine, when administered as follows: mix five or six drops of the tincture in a tumbler half full of water, and to an adult give a table-spoonful at a dose, every hour or two, until relief is obtained.

Administration.—Dissolve six or eight pillets in as many tea-spoonfuls of water, and give a teaspoonful at a dose, and repeat as directed above.

VACCINATION.

We are indebted to Dr. Jenner for the introduction of this artificial inoculation, as a salutary check or prophylactic to one of the most fatal maladies. In the language of Eberle, "there is now no civilized people on earth, amongst whom its blessings have not been largely experienced, and gratefully acknowledged."

DIAGNOSIS AND PROGRESS.—When the human subject becomes inoculated with this *vaccine virus*, it goes through the following course, in order to be a genuine pock. Generally on the third day after the introduction of the matter, a slight inflammation will appear where the puncture was made; on the fourth day it appears more like a pimple, with a faint areola encircling it. The pimple gradually enlarges, until on the latter part of the fifth day, it has assumed a regular circumscribed form, flattened or depressed in the centre, and containing a semi-transparent, limpid fluid. It continues to enlarge from the fifth to the ninth day, at which time it is matured, and some constitutional symptoms appear, such as head-ache, fever, pain, and lameness of the arm, and sometimes an enlargement of the glands of the axillæ.

On the eighth day the areola which first encircled the pock gets darker, and moderately increases until the tenth or eleventh, when it becomes hard, and of a deep red color; the centre of the pustule grows darker, and gradually extends until the pustule is converted into a scab, of a deep mahogany color, and slowly begins to loosen, so as to fall off between the third and fourth week. This is the course and appearance of a genuine pock, and any material deviation from it is spurious. For instance, when the matter forms sooner than the fifth or sixth day, and the areola encircling the pock is of a dark, bluish, or purple hue, the base of the pock hard and indurated, and its centre not depressed, you may infer that it is spurious, and no shield or protection against Small Pox. Therefore, carefully observe the course and appearance of the pustule, in order to be sure of its genuineness.

VARIOLOID. (*Modified Small-Pox.*)

Varioloid is a modified species of Small-Pox, which appears in those who have been exposed to the Small-Pox contagion, after having the susceptibility to taking Small-Pox partially but not fully eradicated by previous inoculation or vaccination.

Diagnosis.—There is not that uniformity in the stages of this variety, as there is in Small-Pox. The eruption generally appears earlier, the pustule matures quicker, and is not filled with opaque, purulent-looking matter; and as the scabs are formed by the drying of the pustules, it is neither so dark nor so thick as in Small-Pox, and instead of leaving pits or marks, by the separation of the scabs, (which usually takes place on the eighth or ninth day,) they leave red discs or rather elevations, instead of depressions, unless some of the scabs are very thick, when the depressions are left, but of a different character from the true variolus mark.

Causes, Treatment, Diet, and Regimen, the same as in Small-Pox.

VEGETATIONS.

W. Acton, in his great work on Venerial Diseases, considers Vegetations under the general term of non-virulent affections, and designates them as Warts, Cauliflower Excrescences, Coxcombs, etc.

When seated on the glans, at the entrance of the urethra in the male or female, or on the inner margin of the prepuce, they are of a vivid red or scarlet color, but when on the skin they are much paler, and sometimes become quite black from exposure. They also differ in consistence and sensibility; they may be quite horny and dry, and possess but little sensibility, or they may be moist and sensitive.

TREATMENT.—*Mercurius, Nit.-acid, Thuja.*

If on administration of the above remedies they do not disappear, touch them occasionally with *Caustic,* and continue the remedies.

DIET AND REGIMEN.—The diet must be light and unirritating, and proper attention paid to cleanliness, by frequent washing with tepid water and Castile soap.

VOICE, LOSS OR SUPPRESSION OF.

A suppression or loss of voice may occur from local as well as general causes; from cold, a disease of the throat or wind-pipe, or from palsy. It has also been known to arise from severe attacks of Small-Pox, Scarlet Fever, and Measles.

TREATMENT.—For loss of voice from palsy, use *Nux-vom., Rhus-tox,* aud *Causticum.*

When there is loss of voice and speech; when the tongue is painful to the touch; the papillæ raised and bright red; and there is dryness of the mouth and throat, use *Belladonna.*

When it depends upon palsy of the tongue, and occurs in damp weather or from cold, use *Dulcamara.*

For loss of speech; the tongue clean, dry, and parched, or palsied, use *Hyosciamus.*

When the loss of speech follows an attack of apoplexy, give *Lachesis* and *Laurocerasus.*

When it depends upon a palsy of the organs of speech; when the face is turgid; circumscribed redness of the cheeks, and a friendly expression, use *Stramonium.*

When there is a complete loss of speech and voice; the breath offensive; or a bad taste; an accumulation of saliva, with a foul, slimy tongue, give *Mercurius.*

ADMINISTRATION.—Repeat the remedy selected every four, six or eight hours, as the case requires, judging from the attack whether it is recent, and with acute symptoms, or chronic.

DIET.—The diet should be nourishing, if the patient is debilitated; but if the disease attacks a person of full habit, it should be restricted in quantity and quality.

WARTS.

These small excrescences, or tuberosities, are sometimes annoying, at least in disfiguring the part.

TREATMENT.—For fleshy or seedy warts, use *Causticum*, one dose every third or fourth day.

When they are flat, hard or brittle, use *Antim. crude*, similarly.

If they are located on the back of the fingers, use *Dulcamara*.

But, when on the sides of the fingers, use *Calcarea-carb*.

When they appear on the face, first give *Causticum* for two or three times as above, then *Dulcamara*, and lastly *Sepia*, in the same manner.

When they appear on the nose, use *Causticum* two or three times a week.

When they are caused by syphilis, use *Mercurius* and *Nitric-acid* in alternation, one dose a day.

ADMINISTRATION.—Give two or three pillets at a dose; or if the medicine is in powders, give as much as will lie on the point of a pen-knife blade.

DIET.—No more attention to diet is necessary, than simply to see that it does not interfere with the treatment.

WHITE SWELLING.

This term is applied to a scrofulous affection of the knee-joint, and is characterized at first by slight pain in the joint after exercise, which gradually increases; the joint begins to enlarge, and the limb becomes emaciated, and debility, night sweats, and symptoms of hectic ensue.

TREATMENT.—Give the first trituration of *Calc.-carb.* and *Ferrum* in two or three grain doses, every four or six hours in alternation. Should the debility increase, and symptoms of hectic appear, such as debility, some afternoon fever, and night-sweats, use *Arsenicum* 1st in alternation with the *Ferrum*, and continue its use, until the symptoms become arrested. Then give *Kali-carb.* 1st, and *Ferrum* in alternation, every six or eight hours, until the patient is restored to health.

EXTERNAL APPLICATIONS.—I have found the wet bandage (wet in cold water,) of very great service in this affection, particularly where there was much local heat.

DIET AND REGIMEN.—The diet ought to be nutritious; such as rare roast beef, rare steak, mutton chop, boiled mutton, etc. The patient should not be allowed to use the limb.

WOUNDS.

Contused wounds are such as are caused by a fall, or blow, or by being pressed between hard bodies.

TREATMENT.—Bathe the part frequently with diluted *Tinct.-arnica*, (one part of *Arnica* to five or six parts of water,) and keep it applied by means of old fine linen wet in it.

LACERATED WOUNDS.

These are caused by tearing, from being caught by machinery, etc.

TREATMENT.—Bring the parts in as close continuity as possible, and keep them so by properly adjusted bandages and compresses, wet in diluted *Tinct.-arnica.* If fever arises, give *Aconite* every three or four hours, until it subsides.

INCISED WOUNDS.

These are produced by a cutting instrument.

TREATMENT.—Bring the parts close together, (in as natural a position as possible,) and secure them thus, by means of adhesive-straps, so as to form a unison by the first intention. Should fever, or inflammatory symptoms set in, use *Aconite* every three or four hours, until they abate.

PUNCTURED WOUNDS

These are produced by a pointed instrument.

They are to be treated by the application of dossils of lint, wet in *Tinct.-arnica*, (diluted as above,) and secured to the part by means of proper bandages.

24*

WORMS.

The origin of intestinal worms is (as we have remarked before) veiled in much obscurity. Some maintain that they are developed from ovula received into the stomach and bowels along with the food and drink; while others assert that they are the result of "*spontaneous generation.*" The latter supposition certainly appears most reasonable, for we know that it is not uncommon to find worms in the bowels of new-born infants. But from whatever source they emanate they are known to be the primary cause of many of the diseases from which children suffer.

TREATMENT.—For the ordinary symptoms of worms, such as variable appetite; foul tongue and offensive breath; distended abdomen; thin and emaciated limbs; irregular bowels; itching of the nose; startings in sleep, and irritability of temper, give *Cina* three times a day. And should inflammatory or feverish symptoms set in after the use of the above remedy, give *Aconite* and *Belladonna*, in alternation, every three or four hours; the latter remedy, particularly if there are any symptoms threatening spasms or convulsions; such as sudden crying, starting, twitching of the muscles, or if the head is hot and the face flushed.

If the abdomen is tumid and hard, and the bowels are constipated, give *Cina* in the morning and *Nux-vom.* at night.

If mucous diarrhœa follows the administration of *Cina*, give *Pulsatilla* first, then *Mercurius*, and lastly *Sulphur*. Of each remedy give two or three doses, that is if the first or second does not have the desired effect.

I prefer the *Cina* in tincture, and generally mix three or four drops in a tumbler half full of pure cold water, and administer as above. The pillets or powder, however, may be used in the same way, by dissolving a small powder or eight or ten pillets.

DIET.—The diet should be restricted to plain, simple food, particularly during treatment.

YELLOW FEVER.

A great diversity of opinion exists in regard to the true character of this form of fever. By some it is considered a typhus, with marked biliary derangement; by others as a congestive bilious remittent; and again some consider it a form of fever entirely independent of either, and dependent upon a specific contagion.

Facts and statistics sufficiently prove that a judicious homœopathic treatment furnishes the only hope of recovery from this most dangerous disease. The signal success that attended the practice of the lamented Taft, during the prevalence of the Yellow Fever at New Orleans, is sufficient to convince the most incredulous, and is but one of many similar instances.

Facts also warrant the assertion that the chances of recovery from Yellow Fever are far greater when the disease is left to the efforts of nature alone than when subjected to the best allopathic treatment ever employed. For when the Yellow Fever prevailed so virulently among the English regiments encamped at Uppark Camp, in Kingston,

Jamaica, the physicians considering it a high grade of Bilious Remittent, prescribed an energetic antiphlogistic treatment, and the result was a most fearful fatality. Every case proved fatal and continued to do so, until the surgeons of the regiments and the resident physicians became appalled at their utter want of success, and discontinued their treatment, to consider among themselves what plan, if any, could be adopted to arrest the disease; and as soon as they ceased to prescribe, the fatality began to subside and the patients to recover. This is not only a matter of history, but I have it from the testimony of Dr. McBean, a physician well known, especially in Jamaica, who was cognizant to all the facts. The same conclusion may be obtained from the report of the Medical Deputation from the French Government to Gibraltar, which furnishes a case of complete recovery, in which the only treatment was a simple bath. Such facts suggest that many others might have terminated as favorably, if left to the efforts of nature alone.

TREATMENT.—The principal remedies for the successful treatment of Yellow Fever are *Ipecac., Aconite, Belladonna, Bryonia, Lobelia, Rhus-tox, Arsenicum*, and *Veratrum*.

When the initiatory symptoms are the following: dizziness, slight chills, an uneasy sensation in the pit of the stomach, nausea, faintness, pain in the back and limbs, languor and general weakness, give *Ipecac.*

But if the commencement of the attack is marked by greater arterial excitement, such as severe pain in the head, flushed face, injected eyes, full or tense pulse, painful uneasy feeling in the stomach, nausea, or nausea and vomiting, give *Aconite.*

When the face is flushed and swollen, the eyes red, the pulse small or contracted, vertigo, dullness, confusion, no power of recognition, hiccough and vomiting of bile, or of bile and mucus, give *Belladonna.*

When the skin becomes yellow, the eyes injected, glassy or suffused with tears, severe burning sensation in the stomach, excessive thirst, pain in the back, general uneasiness, the patient apprehensive and inclined to talk deliriously at night, give *Bryonia.*

When the attack commences with sudden prostration, vertigo, dull, heavy headache, the mouth dry, burning sensation in the throat, incessant nausea, pressure in the pit of the stomach, dark bloody stools, urine of a deep red color, pain in the loins, great weakness and weariness of the limbs, exhaustion and mental despondency, give *Lobelia*.

When there is violent throbbing below the pit of the stomach, nausea, burning eructations, painful distension of the abdomen, pain in the back and limbs, or the limbs are almost paralysed, parched red or brown tongue, the patient gloomy, stupid, and delirious, give *Rhus-tox*.

When the skin is of a reddish-yellow, inclining to brown; the eyes of a dark red, and the vessels congested; coldness of the limbs, with clammy perspiration; the lips dark, the tongue brown or black, the pulse small and tremulous, vomiting of a brownish or blackish substance, followed by great exhaustion, give *Arsenicum* or *Veratrum*, or they may be given in alternation.

ADMINISTRATION.—Mix two or three drops of the tincture, or dissolve ten or twelve pillets in a tumbler one-third full of cold water, and give a tea-spoonful every two or three hours, until the symptoms are mitigated, or another remedy is indicated.

DIET AND REGIMEN.—The diet during the attack should be mild and unirritating; the drinks should consist of cold water and mucilaginous beverages, and the apartments be well ventilated.

INDEX.

	PAGE
Preface	3
List of Remedies	5
Dietetics	9
The Dose	11
Abdomen	16
Abdominal Dropsy	94
Abdominal Pains	17
Abscesses	13
Acne (vide Pimples on the Face)	213
Affections of the Mind	185
After-Pains	17
Ague in the Breast	18
Ague and Fever	124
Ano, Fistula in	127
Apoplexy	19
Appetite, loss of	21
" Voracious	23
Apthæ, or Sprue	25
Asthma	26
Back, Pains in the	199
Biliousness	28
Bilious Colic	30
Bites and Stings of Insects	32
Bite of the Copper-head	32
Bleeding from the Nose	32
" " " Lungs	34
" " " Stomach	37
" " " Urethra	39
" " " Womb	40
Bladder, Inflammation of	154
Bleeding Piles	211
Blindness	42
Boils, or Biles	44
Brain, Inflammation of	171
" Dropsy of the	93

	PAGE
Bronchitis	46
Bubo	48
Burns and Scalds	49
Carbuncle	50
Cataract	51
Catarrh	53
Chancres	55
Chicken Pox	55
Chilblains	56
Cholera	58
Cholera Infantum	61
Cholera Morbus	63
Chronic Bronchitis	46
Chronic Rheumatism	64
Clap	66
Colds	67
Colic	69
Congestion	71
" of the Head	71
" of the Lungs	72
Continued Fever	123
Corns	73
Costiveness	74
Constipation	74
Consumption	76
Contusions	280
Convulsions	77
Cough	79
Cramps in the Limbs	81
" in the Stomach	82
" during Pregnancy	217
Croup	88
Crying of Infants	86
Deafness	141
Delirium Tremens	87
Derangement of the Mind	185

25

INDEX.

	PAGE
Derangement of the Menses.	180
Diabetes	270
Diarrhœa	89
Diseases during Pregnancy.	216
Diseases of the Mind	185
Dislocations	181
Dizziness	90
Dropsy	92
Dropsy of the Abdomen	94
" " Brain	98
" " Chest	101
" " Joints	103
" " Ovaries	105
" " Scrotum	105
Dyspepsia	107
Dysentery	108
Ear-ache	111
Ears, discharges from	112
" Inflammation of	157
Epilepsy	114
Eruptions	115
Erysipelas	117
Eyes, Inflammation of	158
Fainting	119
" during Pregnancy	217
Falling off of the Hair	120
Falling of the Womb	121
Falling of the Vagina	122
Felons	122
Fever and Ague	124
Fever Continued	123
Fever Intermittent	124
Fever Remittent	126
Fever Typhus	266
Fever Yellow	284
Fistula in Ano	127
Fluor Albus	130
Fits, vide Convulsions	77
Fractures and Dislocations	131
Frozen, or Frost-bitten	132
Gathered Breast	133
Gleet	134
Gout	135
Gonorrhœa (Clap)	66
Gravel	188
Hair, falling of the	120
Head-ache	139
Hearing	141

	PAGE
Heart-burn	143
Heart, Inflammation of	160
Heart, Palpitation of	205
Hip-joint Disease	144
Hives	194
Hoarseness	145
Hooping-Cough	147
Hydrophobia	149
Hysterics	151
Inflammation	152
Inflammation of the Brain.	171
" " Bowels.	155
" " Bladder.	154
" " Ear	157
" " Eyes	158
" " Heart	160
" " Kidneys	163
" " Lungs	161
" " Liver	164
" " Nose	166
" " Pleura	214
" " Stomach	167
" " Throat	220
" " Tonsils	220
" " Tongue	168
" " Womb	170
Influenza	172
Irregular Menstruation	180
Itch	173
Jaundice	174
Kidneys, Inflammation of	163
Knee-joint, Swelling of	279
Leucorrhœa	130
Lips, Sore	255
Lock-jaw	175
Loss of Voice	276
Measles	178
Menses Irregular.	180
" Painful	182
" Retention	181
" Suppressed	183
Mental Derangement	185
Milk, Excess of	190
" Depraved quality of	189
" Suppression of	189
Milk Eruption	190
Mind, Diseases of	185

INDEX.

	PAGE
Miscarriage	191
Mumps	192
Nettle Rash	194
Neuralgia	195
Nipples, Sore	255
Nodes	197
Nose, Pains in the	205
Nursing Sore Mouth	198
Pain in the Back	199
" Bowels	69
" Chest	201
" Head	139
" Limbs	222
" Stomach	207
Painful Menstruation	182
Palpitation of the Heart	208
Palsy	210
" of the Limbs	210
Phthisic, vide Asthma	26
Piles	211
Pimples on the Face	213
Pleurisy	214
Pregnancy, Diseases during	216
Prickly Heat	219
Quinsy	220
Remittent Fever	126
Red Gum	221
Rheumatism Acute	222
" Chronic	222
Rickets	225
Ring-Worm	226
Salt Rheum	227
Sea Sickness	230
Seeing	42
Seminal Weakness	231
Scald Head	233
Scarlet Fever	234
Scrofula	238
Scurvy	240
Sickness of the Stomach	241
" During Pregnancy	216
Sight	42

	PAGE
Sleep	243
Small Pox	244
Snuffles	252
Sores, vide Ulcers	267
Sore Eyes	158
" Throat	253
" Lips	255
" Mouth	254
" Mouth while nursing	198
" Nipples	255
Spitting Blood	84
Spots on the Face during Pregnancy	219
Sprains, or Strains	256
Sprue	25
Stye	257
St. Vitus' Dance	258
Squinting	259
Teething	260
Tic Douloureux	203
Tooth-ache	262
Typhus Fever	266
Ulcers	267
Urine, Immoderate flow of	270
" Difficult	272
" Suppression of	271
" Bloody	272
Vaccination	273
Varioloid	274
Vegetations	275
Venereal	66
Voice, Loss of the	276
Voracious Appetite	23
Vision	42
Warts	278
Wetting the Bed	271
Whites, vide Leucorrhœa	130
White Swelling	279
Worms	282
Wounds	281
Yellow Fever	289

THE END

www.ingramcontent.com/pod-product-compliance
Lightning Source LLC
Chambersburg PA
CBHW032059220426
43664CB00008B/1069